Praise for *Digital Bricks and   Property Market*

'Moving home in the UK is a slow, opaque and admin-heavy process for both buyers and property professionals, and the property market is clearly overdue the kind of technology revolution we've seen in other industries. But for those of us who aren't technologists, it's hard to know what that will look like or what it will mean. Reynolds's brilliant book sets out a clear blueprint for the future of the property market, and it's refreshingly easy to read too. I believe it will be a hugely influential book, and it is a must-read for anyone in our industry who wants to know what the future will look like.'

— **Phil Spencer**, property expert and TV personality

'It's very frustrating that nearly a quarter of a century after Rightmove started the UK's property search revolution, the dream – and indeed the expectation – of the digitization of the rest of the UK's home buying and selling process has yet to become a reality. Reynolds's insight and clarity in setting out the intricate web of complex challenges in this area are very well timed to inspire, embolden and enable the necessary stakeholders to come together to finally make that digitization happen. The book is filled with truly inspirational content and vision from someone with an unrivalled breadth of experience in the vital component fields that need to come together, collaborate and act. Reynolds is also disarmingly modest in inviting and promoting constructive discourse. He astutely recognizes that this is a vital component if we are to shift from the tantalizing cusp of a seamless solution to a speedier and better implementation: something that will be of massive benefit to both the industry and millions of future movers.'

— **Miles Shipside**, Co-founder, Rightmove

'The property market in the UK can work well if everyone has the information they need when they need it, and if buyers and sellers pay a fair price for high-quality home moving experts. However, in any one year, transaction levels can fluctuate from 900,000 to 1.5 million, putting huge pressure on the industry as experts take years to train. To provide a transparent twenty-first-century buying and selling process, we must deliver huge amounts of trusted information along with payment systems that work efficiently for everyone: buyer, seller, lender, all the way through to the removal company. Reynolds explains in detail the steps needed to achieve this, including a fascinating history of how the system works now and why it is like it is. He also explains changes such as smart data and AI that will dramatically impact on the property industry, and he shows readers how, through collaboration, change can make a positive difference for all. This is an essential read for those that want to take the necessary steps to deliver the very best service to buyers and sellers in the future.'

— **Kate Faulkner OBE**, leading property expert and commentator

'Buying a home is one of the last consumer experiences to be properly digitized, and in his magnificent book, Reynolds sets out why that is and what the process will look like in the future. It's a bold undertaking, but Reynolds has one of those rare intellects that is capable of distilling huge complexity into breathtaking simplicity. He illuminates the whole landscape and sets out bold new solutions to age-old problems. This is a visionary work: a manifesto for the digitization of the property market, and an invitation to commerce and government to collaborate (or "coadjute", as he likes to say) and make home moving an easier, faster and more transparent process for everyone involved.'

— **Dan Salmons**, CEO, Coadjute

'It's hard to think of an area more in need of digitization than the UK housing market. In *Digital Bricks and Mortar*, John Reynolds provides an excellent overview of the why, the what and the how. He boils it all down into the key insights you really need to understand the space and the urgency, and just how critical it is to the fabric of the UK economy.'
— **Jesper Fogstrup**, former COO and managing director, Compare the Market

'*Digital Bricks and Mortar* is a great resource for anyone interested in the UK housing industry. With its clear, engaging style and visionary insights, this book addresses the current inefficiencies and complexities of home buying, and it outlines very clearly how technology (including distributed ledger mechanisms) can transform the process. The author underscores the importance of user experience and the need for a customer-centric approach throughout. The book turns complex digital concepts into practical, actionable strategies, relevant to buyers, sellers, agents and legal professionals. It's a must-read for professionals eager to streamline their operations and enhance transparency. The included case studies also serve to alleviate fears and scepticism by showing that successful digital transformations are not just possible, but have already been achieved elsewhere.'
— **Elizabeth G. Chambers**, fintech investor and board director at TSB Bank, Currensea and Wise plc

'John Reynolds starts by reminding us of the innovations we have collectively delivered throughout history, and then, having set this context, he helps us see into the next phases of innovation in property and money, enabling their digitization and improved interoperability. I thoroughly enjoyed Reynolds's insights and approach, and I recommend that others who want to be among the first to understand the next steps in this digitization journey read this book. I am very excited to work with Reynolds and fellow readers, innovators and collaborators to bring some of this vision to life.'
— **Peter Left**, digital and markets innovation leader at Lloyds Banking Group

'Many scale-ups have tried and failed to "solve" the UK property market's challenges. Praetura have supported Reynolds and the Coadjute team because they think of the housing market's challenges on a macro scale, working with incumbents and not against them. Reynolds's knowledge of digital property has been integral to bringing this dream to life. This book is a must-read for anyone looking to understand more about the power technological innovation can have in this space and what is practically required to get there.'

— **Andy Barrow**, Partner, Praetura Ventures

'Providing invaluable insights into the future of home buying and digitization, this is a must-read for anyone working in the UK property industry. Despite being one of the most important purchases of our life, the home-buying experience is fraught with complexity. The gaps and what is needed to address them are laid bare in this book, with clear explanations and context in an increasingly digital world. We've made exceptional progress on collaboration, digitization and interoperability, but the vision can only be realized with the commitment and enterprise of industry visionaries like Reynolds.'

— **Maria Harris**, founder and chair of the Open Property Data Association

'Despite what many people think, the practice of conveyancing – and of the home moving process more generally – is complex and multilayered. It therefore takes confidence and a deep understanding of how change can be made if we are to avoid undermining public faith in the systems as they currently operate. Reynolds is to be congratulated on producing a considered and very thought-provoking analysis – one that will hopefully result in real and long-lasting change for the benefit of consumers and those involved in advising them when buying and selling property.'

— **Stewart Brymer OBE**, honorary professor in law at the University of Dundee and conveyancing expert

Digital Bricks and Mortar

Dear Hugh,

Thank you for all you do for Coadjute.

Hope you enjoy my book.

Best wishes
John

PERSPECTIVES ON BUSINESS

Series editor: Professor Diane Coyle

Why You Dread Work: What's Going Wrong in Your Workplace and How to Fix It — Helen Holmes

Digital Transformation at Scale: Why the Strategy Is Delivery (Second Edition) — Andrew Greenway, Ben Terrett, Mike Bracken and Tom Loosemore

The Service Organization: How to Deliver and Lead Successful Services, Sustainably — Kate Tarling

Money in the Metaverse: Digital Assets, Online Identities, Spatial Computing and Why Virtual Worlds Mean Real Business — David Birch and Victoria Richardson

Digital Bricks and Mortar: Transforming the UK Property Market — John Reynolds

Digital Bricks and Mortar
Transforming the Property Market

John Reynolds

LONDON PUBLISHING PARTNERSHIP

Copyright © 2024 John Reynolds

Published by London Publishing Partnership
www.londonpublishingpartnership.co.uk

Published in association with
Enlightenment Economics
www.enlightenmenteconomics.com

All Rights Reserved

ISBN: 978-1-916749-08-5 (pbk)
ISBN: 978-1-916749-09-2 (iPDF)
ISBN: 978-1-916749-10-8 (epub)

A catalogue record for this book is
available from the British Library

This book has been composed in Candara

Copy-edited and typeset by
T&T Productions Ltd, London
www.tandtproductions.com

Printed and bound in Great Britain
by Hobbs the Printers Ltd

Contents

Foreword by Johan Svanstrom, CEO, Rightmove   xiii
Preface   xv

PART I
THE BIRTH OF TODAY'S MARKET   1

CHAPTER 1
The early market   3

CHAPTER 2
Market growth   9

CHAPTER 3
The arrival of the internet   23

CHAPTER 4
The online market is broken   39

PART II
DIGITAL FOUNDATIONS   49

CHAPTER 5
Market infrastructure   51

CHAPTER 6
Data standards   69

CHAPTER 7
Digital trust: the golden thread   83

CHAPTER 8
Digital identity   101

PART III
DIGITAL MONEY AND TITLES   117

CHAPTER 9
Tokenization   119

CHAPTER 10
Digital money 133
CHAPTER 11
Digital titles 149

PART IV
THE DIGITAL PROPERTY MARKET 161
CHAPTER 12
A housing super app 163
CHAPTER 13
The business ecosystem 177
CHAPTER 14
Digital completion 187
CHAPTER 15
Smart contracts 201

PART V
THE INTELLIGENT MARKET AND THE PROPERTYVERSE 213
CHAPTER 16
Artificial intelligence 215
CHAPTER 17
The intelligent market 221
CHAPTER 18
The propertyverse 235
CHAPTER 19
Market supervision 247

Epilogue 261
Acknowledgements 263
About the author 265
Endnotes 267

This book is dedicated to my parents, Pat and Bridie, for their love, support and prayers. And to my boys, Lewis, Liam, Frank and Seamus – my love for you has been my greatest source of motivation to dig deep and always keep pushing.

Foreword

The potential to vastly improve the property buying and selling process in the UK is an exciting one. Better yet, the recent digital innovation in this space makes it highly likely there will be transformational change over the next five to ten years.

In the UK we are miles behind some of our European counterparts in the length of time it takes to move home. Even in 2024, if you speak to the majority of people who have completed a UK property transaction, you will hear that it was frustratingly long, with so much paperwork and many disconnected parties.

Our data shows that on average it is currently taking a painful seven months from someone putting their property up for sale on Rightmove until they pick up the keys. Seven months means that if they want to be in their new home before Christmas then they need to be ready to sell at Easter!

But change is coming. As someone who has spent over two decades in the technology sector, I have seen first hand the transformative power of digital innovation. I have witnessed how the fusion of streamlined processes and cutting-edge technology can revolutionize entire industries, creating end-to-end journeys that are efficient, transparent and user-friendly. And I believe that this same transformation is not only possible but inevitable in the property market.

This is where *Digital Bricks and Mortar* comes in. In this book John Reynolds offers a comprehensive and compelling vision of what a truly digital property market could look like. Drawing on his deep expertise, Reynolds maps out the key elements required to build this market, from the foundational data standards and digital identities to the game-changing potential of blockchain, artificial intelligence and the metaverse.

What makes this book so valuable is that it does not just focus on the here and now. Reynolds takes a long-term, forward-looking view, exploring how emerging technologies could reshape the property market in ways we can barely imagine today. His insights into the future of digital currencies, smart contracts and immersive technologies offer a tantalizing glimpse of a world where the process of buying and selling property is seamless.

But for all its visionary ideas, *Digital Bricks and Mortar* remains grounded in the practical realities of the market. Reynolds draws on his experience leading cutting-edge projects with major players such as HM Land Registry and the Bank of England to show how his ideas can be translated into action. He also stresses the crucial importance of collaboration and shared standards, highlighting the excellent work already being done by industry groups and associations.

Whether you are an estate agent, a mortgage lender, a conveyancer, a policymaker or a tech innovator, there are many valuable insights and ideas in here that will inform your thinking and shape your strategies for years to come. It is a roadmap for building a property market that is truly fit for the digital age – a market that is transparent and efficient and always puts the customer first.

The journey ahead will not be easy. Transforming a market as complex as property will require vision, collaboration and hard work from all of us. But I am excited and energized by the vision John Reynolds presents in these pages.

I believe that by embracing the ideas and initiatives he explores, together, we can build a property market that is a model for the rest of the world – a market that harnesses the full potential of the digital age to deliver faster, safer and more certain transactions for all.

Johan Svanstrom, CEO, Rightmove

Preface

There is a widespread consensus that the UK property market fails to serve the needs of both consumers and the businesses that cater to them. By many measures it is one of the worst-performing property markets in the Western world. Particularly striking are indicators such as the length of time required for a transaction, the high proportion of transactions that fall through and the stress experienced by consumers. However, in recent years a promising solution has emerged: a digital property market. This concept has been heralded by the UK Digital Property Market Steering Group as a panacea, offering a vision of a future characterized by smoother, faster and more certain transactions.[1] According to HM Land Registry's strategy document for 2022 onwards, this digital renaissance will create 'a world-leading property market as part of a thriving economy and a sustainable future'.[2]

Despite the high expectations placed on digitalization to resolve the UK's dysfunctional property market, there has been a surprising lack of literature addressing the substance of this future market. Key questions remain unanswered. For instance, what exactly is the digital property market? Who is responsible for building it, and who will fund its development? Where will the capital investment come from? Who will supervise it, and how? Are we following a blueprint from some other market, and if so, which one?

Current conversations about the digital property market focus on topics such as data standards, digital identity, trust frameworks and upfront information, but there is a lack of meaningful literature on how these things will manifest and be implemented. There is also limited discussion on the rapidly approaching wave of innovation that will be triggered by digital assets, artificial intelligence (AI) and the metaverse, or as I refer to it in this book, the propertyverse.

This book aims to fill these critical gaps by comprehensively documenting the current industry discourse, connecting the dots between the various elements, and expanding the conversation to include a more forward-looking examination of innovative technologies and their disruptive implications for the UK property market.

I have spent thirty years in technology transformation and the better part of the last decade in the property market. As COO and co-founder of Coadjute, a transaction platform for the property market, I have, together with its other co-founders, grown the company to a point where it has recently secured strategic investment from the UK's largest property portal, Rightmove, and the country's three largest mortgage lenders: Lloyds, Nationwide and NatWest.[*]

Over recent years I have designed and led some of the industry's most innovative projects, including HM Land Registry's exploration of blockchain and tokenization; Project Meridian, an experiment by the Bank of England and the Bank for International Settlements in which we proved the benefits of real-time movement of funds; and, most recently, work with both Mastercard and the Regulated Liability Network on property transaction settlement using tokenized commercial bank money.

As well as leading a range of innovation projects, I have spent a significant amount of time working on industry collaboration. I was, for example, one of the founding members of the Home Buying and Selling Group's data standards working group and the UK Open Property Data Association. I have also worked with

---

[*] It is important to note that everything in this book reflects my personal opinions and does not necessarily represent the views of my company or my investors. No information provided in this book is commercial or confidential.

the Department for Culture, Media and Sport on alpha testing the digital identity and attributes trust framework, and with LawtechUK on their smart legal contract research. Through all of this – from designing and leading the build of the Coadjute platform using enterprise-grade blockchain technology, to crafting application programming interfaces (APIs) and our current work on AI, to the connections and interactions with our customers and stakeholders – I have learned a lot about the digital property market.

Throughout this book I draw on the experiences and insights I have gained from working on cutting-edge projects and from collaborating with key stakeholders across the industry. My goal is not to provide definitive answers or prescriptive solutions, but rather to document this complex market and create a reference manual for those working in, or entering, the market. I want to share my hard-earned knowledge so that others might pick it up, learn from it, innovate and build on it in order to improve the market. I believe that by working together and building on the ideas and initiatives explored in this book, we can create a digital property market that benefits all stakeholders and provides the digital market that consumers desire and deserve.

This book is for property professionals, mortgage experts, software vendors, policymakers, regulators, academics and all other property market stakeholders. Regardless of your role or perspective, this book should provide valuable insights and ideas that can inform your thinking and decision making.

How to read this book

This book is structured in four parts, each designed to provide a comprehensive understanding of the digital property market and its future potential.

The first part, 'The birth of today's market', gives a brief history of the property market, taking us from the Domesday

Book up to the emergence of the internet. This part sets the stage for understanding the current state of the market and the challenges it faces.

The second part, 'Digital foundations', sets out in detail the key topics being explored and championed by the Digital Property Market Steering Group. Here I provide an in-depth analysis of the essential building blocks needed to create a robust and efficient digital property market, such as data standards, digital identity, market infrastructure, trust frameworks and upfront information.

In the third part, 'Digital money and titles', I explore the exciting world of innovation in digital currencies and the potential for digital titles and digital completion, available 24 hours a day, 365 days a year. This part of the book examines how current digital currency innovations can revolutionize the way property transactions are conducted, making them faster, more secure and more transparent.

The book concludes with a part titled 'The intelligent market', which looks at the transformative potential of AI and immersive technologies in the property market. This part paints a vivid picture of how these technologies can reshape the way we interact with property, creating new opportunities for buyers, sellers and professionals alike.

While each part and each chapter of this book is designed to be largely self-contained, allowing you to read them as stand-alone pieces, I highly recommend reading the book from beginning to end. By doing so, you will gain a comprehensive understanding of the digital property market, its foundations, the role of digital assets and the exciting possibilities that lie ahead in the intelligent market and the propertyverse.

So, whether you are a seasoned property market professional looking to stay ahead of the curve, a software provider seeking to develop game-changing solutions, an organization aiming to drive industry-wide change, a policymaker shaping the regulatory environment or simply someone with a keen interest in the future of property transactions, this book should have some valuable insights for you.

For all property market stakeholders, let us envision and create a future in which the digital property market is a reality – a reality in which buying a home is a seamless experience; where people can find their perfect place with ease; where transactions are safe, quick and transparent; and where the process of buying and selling property is as smooth and enjoyable as it can be. Ultimately, it is about creating the kind of home-buying experience that consumers expect and deserve: one that is empowered by digital technology but never loses sight of the human element at the heart of every transaction.

PART I

THE BIRTH OF TODAY'S MARKET

The property market had a rich and wonderful history before digital technology arrived, and the activities at its core – exchanging money for deeds and recording the transaction – are not, in themselves, complex things to do.

CHAPTER 1

The early market

While at first glance this chapter may seem unrelated to the digital property market, understanding the history provides valuable insights into today's market dynamics. Below I lay out the key moments and developments.

Our journey into the market's history begins with the Domesday Book of 1086, just before William the Conqueror's death. This revolutionary survey transformed England's taxation and governance, listing lands worth approximately £73,000 and providing William with an annual income of roughly £22,500, making him one of the wealthiest individuals ever in today's terms.

The Domesday Book was the first comprehensive document to offer a profound understanding of English society at that time. It revealed a significant shift in land ownership and covered the destruction caused by William's conquest, marking the beginning of a structured approach to documenting property ownership. This effort introduced a level of record-keeping detail that remains a hallmark of modern land tenure systems.

Feudalism post-conquest introduced a hierarchy of land ownership and duties, significantly influencing property rights and responsibilities. The Magna Carta, signed in 1215, was pivotal in limiting the king's power and protecting landowners' rights, laying early groundwork for modern legal systems around property.

The Black Death in the middle of the fourteenth century led to labour shortages and shifts in land value and ownership,

3

altering the economy. The Hundred Years' War (1337–1453) and the Wars of the Roses (1455–85) further destabilized land ownership, leading to significant land confiscation and redistribution.

The Tudor era, starting with Henry VII's reign in 1485, marked another transformative period. The Statute of Uses, introduced in 1535, aimed to simplify the conveyance process of land ownership and ensure the Crown collected its dues. The Act addressed the use of trusts to circumvent feudal dues and taxes, making property transfer more straightforward. Later, the Bill of Rights, passed in 1689, would protect property rights by prohibiting the monarch from interfering with the law and imposing taxes without parliament's consent.

The establishment of the Bank of England in 1694 marked a pivotal moment in financial history. Initially serving as a private entity to support the government's financial requirements, the bank was granted the right to issue banknotes in exchange for raising funds for William III's war against France. In 1725 the bank introduced denominated banknotes, laying the groundwork for modern paper currency. By 1734 it had settled into its current home in the City, earning the nickname 'the Old Lady of Threadneedle Street'.

During the French invasion scare in 1797, the public rushed to convert banknotes into gold, depleting the bank's reserves. To protect the remaining reserves, the prime minister ordered the bank to stop paying notes in gold, leading to the famous cartoon depicting William Pitt the Younger attempting to 'woo' the gold off an old lady representing the bank.

The Industrial Revolution saw the rise of private banks, with the Royal Bank of Scotland and Lloyds Bank emerging as prominent players. However, the Bank Charter Act of 1844 limited private banks' note issuance, granting the Bank of England a monopoly.

Building societies, such as Ketley's Building Society and the Halifax Permanent Building Society, also emerged during this period. These institutions aimed to support families facing adversity and provided a means for members to collectively contribute funds and borrow for home construction or purchases. The Regulation of Benefit Building Societies Act of 1836 marked the first official recognition of building societies, promoting savings within these institutions.

Before the eighteenth century the concept of mortgages was not fully developed. Wealthy landowners and aristocrats used land as collateral for loans in mostly informal and private agreements. The emergence of building societies during the nineteenth century laid the foundations for modern mortgage lenders. Legal reforms standardized mortgages and made them more accessible, reflecting the growing concept of home ownership among the middle classes.

World War I led to a housing shortage and a halt in house building, affecting the mortgage market. The UK government took a more active role in housing after the war, increasing the availability of affordable housing and supporting mortgage lending. The 1920s housing boom, partly fuelled by government initiatives, expanded suburbs and increased mortgage availability for average workers. By 1925 building societies were the main mortgage providers, setting the stage for further expansion and the evolution of modern mortgage systems.

The roots of modern insurance companies can be traced back to the property market, with the Great Fire of 1666 being a pivotal event. The first fire insurance company, the Insurance Office for Houses, was established in 1681 by economist Nicholas Barbon and his associates. The Sun Fire Office, later known as Sun Insurance, was founded in 1710 and then expanded across the country. By the 1790s it was a dominant player in the fire insurance sector.

Alliance Assurance emerged in 1824, and the Royal Insurance Company was established in 1845, with both expanding rapidly. In 1959 Alliance Assurance merged with Sun Insurance to form Sun Alliance, which later joined forces with Royal Insurance, resulting in Royal and Sun Alliance, now known as the RSA Insurance Group. The Legal & General Life Assurance Society, founded in 1836, played a significant role in the British life insurance industry and also diversified into lending and property.

The history of British estate agencies dates back to a time before the profession existed. Early estate agents emerged from professions such as surveying and auctioneering, with selling houses being a secondary activity. The nineteenth century saw a growth in home ownership alongside the rise of building societies, leading to the emergence of estate agents to facilitate property transactions. Notable agencies such as Strutt & Parker and Savill & Son – initially family businesses – were born out of this era.

In 1868 the Royal Institution of Chartered Surveyors (RICS) was established in London, marking a significant milestone for the surveying profession. The awarding of a Royal Charter in 1881 signified RICS's critical role in advancing the profession for public benefit, both in the UK and internationally. RICS was at the forefront of guiding the development, planning and maintenance of cities and towns, influencing the lives of generations and the shape of the built environment.

The Statute of Frauds, enacted in 1677, mandated that certain contracts, including land sales, must be in writing and signed by all parties, setting a precedent for preventing fraud and resolving disputes. From that point the transfer of property became a heavily legal and contractual process, in which the Law Society would play a significant role. The Law Society's origins can be traced back to the eighteenth century, with its earliest form being the Society of Gentleman Practisers in the Courts of Law and Equity, founded in 1739. The Law Society as we know it today was revived by Bryan Holme as the London Law Institution, eventually becoming the Incorporated Law Society. The society acquired its Chancery Lane site in 1827 and was granted a Royal Charter in 1831. The Law Society played a crucial role in legal reforms and education, evolving into a comprehensive professional organization serving the legal profession.

Due to the difficulty of proving land ownership for securing loans, particularly in Yorkshire's cloth manufacturing industry, an Act of parliament in 1704 established public registries in Yorkshire and Middlesex. The Wakefield Registry of Deeds, inaugurated in 1704, was the first of its kind, and it was followed by registries in the East and North Ridings of Yorkshire. These registries recorded legal documents concerning land ownership, although registration was not mandatory.

It is interesting to read the Land Registration Act of 1862 as it shows that real-time, immediate transactions and registry updates were possible as early as the middle of the nineteenth century, with parties and their agents attending the Land Registry office and completing the transaction then and there. The Act introduced a more robust process for land registration that included actual land titles in its records, providing official proof of ownership. The first registration under the Act occurred on 24 June 1863, with the Chantry in Sproughton, Suffolk, registered under title number 1.

Technology began to play a role around this time. Invented by Alexander Graham Bell in 1876, the telephone was revolutionary, enabling direct voice communication over long distances. The telephone was introduced to Britain in 1878, leading to the formation of the Telephone Company Ltd. The Post Office was initially sceptical but eventually acquired telephone companies to create a national network. The telephone transformed property transactions by breaking down geographical barriers and facilitating a new era of connectivity.

The early twentieth century brought key legislation. Post-World War I, the British government introduced the 1919 Housing and Town Planning Act – or the Addison Act – to address the housing crisis and provide healthy living spaces. The Act marked the beginning of direct government funding for housing construction, with local councils planning and developing new housing schemes.

The Law of Property Act of 1925 aimed to simplify and consolidate property law, reducing the types of legal estate in the land to freehold and leasehold, standardizing the conveyancing process and facilitating property transfer. The Act also made provisions for protecting tenants' rights.

The Land Registration Act of 1925 reformed the land registration system in England and Wales, encouraging the registration of land ownership and interests. The Act established a comprehensive and compulsory system for registering land titles that adhered to the 'mirror', 'curtain' and 'insurance' principles and that made property transactions more secure and predictable.

*

By 1925 the key elements of the modern property market were solidly established, enabling its smooth functioning.

This structure now encompasses HM Land Registry, created to document property ownership; the Bank of England, which has a significant role in steering the economy; and major mortgage providers such as Lloyds, Halifax, Nationwide and NatWest, which supply essential financing for property acquisitions. Insurance giants such as the RSA Insurance Group offer protection against various risks, while renowned estate agents such as Chestertons and Savills facilitate property transactions. The surveying profession, guided by RICS, ensures property valuation and standards, while the Law Society supervises legal expertise within the sector.

These elements are underpinned by a legislative framework that includes significant Acts such as the Banking Charter Act, the Building Societies Act, the Law of Property Act and the Land Registration Act, collectively forming the legal foundation of the contemporary property market.

This infrastructure enables the effective operation and expansion of the property market, demonstrating a system that has adapted to cater to the intricate needs of buyers, sellers and industry professionals.

CHAPTER 2

Market growth

In this chapter I cover the twentieth-century evolution of the UK property market up to the mid 1990s, highlighting its growing complexity and expansion. I start with the traditional physical market, where transactions were tangible and conducted in person, showcasing an era of direct engagement and real-time updates at land registry offices. As you read about this, it is worth considering the market's inherent efficiency and potential for simple and certain transactions to be conducted in real time.

I also cover the key socio-economic impacts, from the Great Depression to post-World War II reconstruction, illustrating how these events shaped housing policies, market dynamics and financing innovations. I cover the transformations brought about by slum clearances, the Right to Buy scheme and the role of technology in streamlining processes, emphasizing the market's adaptability to legislative changes and societal needs.

I aim to bring to life how significant legislative and technological shifts – such as deregulation and the introduction of computers during the 1980s – have the power to radically transform the market, just as the internet would do a decade after the Thatcher era, and AI has started to do today.

Simpler, safer and more certain transactions

In the era before the internet, property transactions were characterized by a level of simplicity, certainty and tangibility that made the process straightforward for both buyers and sellers. The physical nature of these transactions created an inherent trust and understanding among all parties involved.

In the traditional physical property market, buyers and sellers interacted directly with the service providers integral to their transactions. Conveyancers, lawyers and banks worked closely with their clients, often meeting face-to-face to discuss the details of the transaction. This personal interaction fostered a sense of trust and transparency, as all parties were able to communicate directly and address any concerns or questions in real time.

One of the most significant aspects of the physical property market was the tangible nature of the transaction itself. Title deeds, mortgage documents and other legal papers were physical documents that could be held, signed and exchanged in person. This tangibility provided a sense of security and certainty for both buyers and sellers, as they could see and touch the documents that represented the transfer of ownership.

Similarly, monetary exchanges were also physical in nature. Buyers would provide certified cheques or bank drafts to sellers, creating a clear and tangible record of the financial transaction. This physical exchange of funds added to the overall sense of security and certainty in the transaction, as both parties could see and verify the transfer of money in real time.

Another critical aspect of the physical property market was the ability of individuals to carry their identity from one physical location to another seamlessly. In face-to-face interactions, buyers and sellers could easily verify each other's identities, reducing the risk of fraud or misrepresentation. This inherent trust in the physical identity of the parties involved further contributed to the simplicity and certainty of the transaction.

During this period, a notable practice outlined in the 1862 Land Registration Act allowed parties and their agents to convene at the Land Registry office to execute transactions and update the register 'then and there'. This ability to gather all necessary parties in one physical location and complete the transaction in a single sitting was a testament to the efficiency and immediacy of the physical property market. By bringing together buyers, sellers, conveyancers and Land Registry officials in one place, the process of transferring ownership and updating official records was streamlined and simplified.

The historical ability to exchange information and update property records in real time presents a compelling model for today's digital property market, and this is a theme we will return to throughout the book.

The Great Depression (1929–39)

The Wall Street Crash of 1929 initiated the Great Depression, profoundly impacting the UK housing and mortgage markets amid a global economic downturn that stretched from 1929 to 1939. This period saw a dramatic decline in property values and construction, coupled with a tightened mortgage market as lenders faced the increased risk of defaults due to soaring unemployment and economic instability.

In the UK the Depression exacerbated the existing housing crisis, especially in industrial regions reliant on sectors such as coal and steel, where job losses and reduced incomes made home ownership and mortgage maintenance increasingly difficult. The result was a marked decrease in home ownership rates and a spike in foreclosures, as financial uncertainty made it challenging for many to secure or sustain home loans.

Government responses, characterized by austerity measures, were initially ineffective at mitigating the housing market downturn. The situation began to improve in the late 1930s, however, thanks in part to the devaluation of the pound and lower interest rates, which somewhat revived the housing market by making British exports – and by extension, the broader economy – more competitive. This slow recovery saw a gradual increase in construction and a slight easing in mortgage lending, though recovery rates varied significantly across the country.

Post-World War II reconstruction (1945–51)

The post-World War II era, governed by Prime Minister Clement Attlee's Labour administration, was a pivotal time for the UK housing market. The war had resulted in a critical shortage of homes, necessitating urgent and innovative responses. In response to the shortage, the government embarked on a mission to provide affordable housing for all social classes. This led to the introduction of prefabricated houses and modern construction techniques, expediting the house-building process.

Building societies – both small regional players and large societies such as Halifax and Nationwide – played an instrumental role during this period. Their mortgage finance options were vital in supporting the housing boom. These developments, alongside the implementation of the New Towns Act of 1946,[1] facilitated the construction of more than 800,000 homes and the development of new towns such as Stevenage, Milton Keynes and Harlow. This Act was critical in alleviating overcrowding in cities by redistributing population growth.

The Town and Country Planning Act of 1947 was another landmark in UK urban planning. It aimed to balance the needs of property owners, developers and the community, promoting orderly and sustainable development. The Act also enabled slum clearances, leading to the redevelopment of areas with substandard housing and the creation of modern, sanitary living spaces.

Slum clearances and the evolution of the right to buy (1951–64)

The UK housing market continued to evolve between 1951 and 1964 under the Conservative governments led by Winston

Churchill, Sir Anthony Eden, Harold Macmillan and Sir Alec Douglas-Home. The era was marked by extensive suburban housing development.

While the slum clearance initiatives from the 1930s to the 1970s improved living conditions, they often failed to benefit the poorest citizens.[2] Cities such as Liverpool and Manchester demolished numerous slums, but the new housing was frequently unaffordable for the displaced, and the upper working class were instead often the ones to benefit.

Despite improving the quality of housing, slum clearance programmes often failed to cater to the needs of the most impoverished, disrupting communities and affecting millions. These projects, exemplified by the Byker Wall in Newcastle, disproportionately impacted vulnerable populations by displacing them from their homes and breaking up crucial social support networks. Inadequate community consultation and support for those who were relocated exacerbated the negative consequences. As a result, only a small fraction of the original residents, and an even smaller fraction of the most vulnerable, were able to return to areas such as the Byker Wall once the redevelopment was complete. The failure to prioritize the needs of disadvantaged residents in these programmes led to significant social upheaval and long-lasting negative impacts for the affected communities.[3]

The House Purchase and Housing Act of 1959 removed the requirement of ministerial consent for the sale of council houses. However, tenants did not have a default right to buy as they still required an agreement from the local authority to purchase their house.

Urban challenges and economic turmoil (1964–79)

Between 1964 and 1979 the UK grappled with urban difficulties and economic instability. Under the Labour prime minister Harold Wilson and subsequent governments, the focus shifted towards addressing urban decay and inadequate housing. The governments' strategies included the construction of council tower blocks, offering affordable housing solutions.[4] However, these developments later encountered maintenance challenges and created social issues.

The 1970s were marked by economic hardships, such as high inflation and rising interest rates, adversely affecting property values and affordability. The property market during this period was characterized by its instability and fluctuating values.

Decimalization (1970–9)

The switch to decimal currency in 1971, known as Decimal Day, had significant implications for the banking and property sectors. Under Edward Heath's Conservative government, and later under Wilson's Labour government, this transition required substantial changes, including updating financial systems, recalibrating ATMs (automated teller machines) and retraining staff.

But decimalization also simplified property transactions. Pricing, mortgage calculations and rents became more transparent and easier to comprehend. This clarity in financial dealings likely fostered greater participation in the property market by reducing the complexity of financial understanding.

The Thatcher era (1979–90)

In 1980 the UK government enacted the Housing Act, enabling local authority tenants to buy their homes at a discount, thereby fulfilling a Conservative Party manifesto pledge. The Right to Buy scheme, which was also applied in Scotland and Northern Ireland with slight variations, required a minimum of three years' tenancy.[5] Discounts started at 33% and increased with tenancy length, being capped at £50,000 – now equivalent to more than £200,000. However, these discounts were repayable if homes were sold within five years. Notably, the Act did not oblige councils to allocate the sales revenue to new social housing construction. Two key pieces of legislation played a crucial role during this period. The Financial Services Act of 1986 established a regulatory framework that later encompassed mortgage broking. This Act represented a significant step for consumer protection in a complex financial market. The Building Societies Act of 1986 allowed these institutions to expand their services and directly compete with banks. This era introduced diverse mortgage products, such as fixed-rate and adjustable-rate mortgages, offering consumers more choices. The deregulation of mortgage interest rates led to a more competitive lending environment.

The 1980s saw significant financial fluctuations in the housing market. Initially, the Right to Buy scheme led to a surge in property transactions and home ownership, causing a sharp increase in property values. However, a downturn in the mid 1980s, driven by rising interest rates and economic challenges, led to a housing market crash and widespread negative equity, highlighting the market's sensitivity to economic conditions.[6]

Mergers and acquisitions

The late 1980s and early 1990s saw significant mergers and acquisitions in the banking sector. A notable example was the merger of Lloyds Bank with TSB Group in 1995, forming Lloyds TSB. Many building societies chose to demutualize and become banks, seeking wider financial opportunities and market access. Halifax was the first to do so, taking advantage of the 1986 Act. Nationwide, however, remained a building society, as did some smaller entities.[7]

Home ownership growth

Margaret Thatcher's policies, particularly the Housing Act of 1980, coupled with a diverse range of mortgage products and increased lender competition, led to a dramatic increase in home ownership rates. This era witnessed the sale of millions of council houses and a significant growth in the proportion of homeowners in the UK. At the same time, the Thatcher government ended the ability of local authorities to build housing, and although housing associations emerged to provide social housing, the annual volume of home construction in the UK collapsed and has never regained its pre-1980 levels. These policies and the subsequent cycles of boom and bust in the housing market have profoundly influenced the UK's property landscape, leaving a lasting impact on housing trends and ownership patterns.

Mortgage brokers

In the 1970s and 1980s the UK mortgage market underwent a significant transformation. This era marked a shift from a landscape traditionally dominated by banks and building societies to an increasingly diverse market featuring independent mortgage brokers. This change was gradual and organic, emerging as a response to the evolving financial services market and the changing needs of consumers.

The establishment of John Charcol in 1974 symbolizes this pivotal shift. As one of the first independent mortgage brokers in the UK, John Charcol not only introduced a novel approach to accessing mortgage products but also significantly influenced consumer attitudes towards property financing.

In the late 1980s London & Country Mortgages emerged, exemplifying the trend towards offering fee-free mortgage advice. This approach was customer-centric and challenged the traditional models of mortgage advising. Alexander Hall Associates, established in 1987, further showcased the innovation and dynamism within this nascent industry.

The late 1980s witnessed a rapid increase in the number of mortgage brokers, accompanied by a greater variety of mortgage products. And the growth was not only in terms of numbers: there was also a move towards higher professional standards in the industry. This shift happened in a context where specific regulatory standards for mortgage broking were still evolving.

The early 1990s saw the entry into the sector of Savills Private Finance, focusing on the high-end market. This development underscored the adaptability and breadth of the burgeoning mortgage broking sector.

Estate agents

The immediate post-war years were focused on rebuilding the UK's housing stock, which had been severely damaged during World War II. The government initiated large-scale construction projects, leading to an increased demand for estate agency services.

The 1960s and 1970s were also characterized by a housing boom, this time driven by economic growth and increased consumer spending power. This period saw the rise of home ownership as a key aspect of the British dream, boosting the estate agency market further. Prominent estate agency chains began to emerge during this time, expanding their presence across the UK. Brands such as Foxtons (founded in 1981, but with roots going back to the late 1970s) and Savills and Knight Frank (which both had histories predating this period but which now saw significant expansion) became well-known names. Winkworth, founded in 1835, also began franchising in the late 1960s, contributing to the trend of brand expansion.

The Estate Agency Act of 1979 was a significant piece of legislation introduced towards the end of this period. It aimed to professionalize the industry by setting minimum standards for estate agents, and it provided consumers with protection from unfair practices.

While the industry in the late 1970s primarily relied on traditional forms of advertising such as newspaper listings, the groundwork for future technological integration in property listings and management was already being laid. Estate agencies began to employ more sophisticated marketing techniques, including the use of branded company cars, detailed property brochures and the opening of high street shops designed to attract walk-in customers.

The UK economy experienced several recessions during this period, notably in the mid 1970s, and these impacted the property market. High inflation rates and unemployment affected consumer confidence and, by extension, the estate agency business.

Licensed conveyancers

In the post-World War II period, conveyancing in the UK was largely controlled by solicitors. This era was characterized by traditional, manual processes, with solicitor firms embedded in local communities. The conveyancing market during this time was relatively stagnant, seeing minimal competition or innovation.

A transformation in conveyancing accompanied the Thatcher government's free-market reforms in the 1980s. Known for its commitment to deregulation and market liberalization, Thatcher's Conservative government identified the legal services market, including conveyancing, as a target for reform.

The passing of the Administration of Justice Act in 1985 was a landmark event. This legislation, aligned with Thatcher's liberalization policies, permitted non-solicitors to qualify as licensed conveyancers, breaking solicitors' long-held monopoly in the conveyancing market. This change was partly in response to public dissatisfaction with the high costs and inefficiencies of traditional conveyancing services.[8] The Act led to the creation of the Council for Licensed Conveyancers, which regulated this new profession. These developments introduced competition into the market, providing consumers with more options and often more competitive pricing, and significantly altering the market for conveyancing services in the UK.

During the 1970s and 1980s the UK conveyancing industry adopted new technologies, integrating computerized systems to streamline processes. This period saw a transition from traditional, manual methods to more efficient, technology-driven ones, but the full impact of these technologies was still on the horizon. The housing market experienced a price boom in the late 1980s – one that pushed the limits of conventional conveyancing methods. The surge in transaction volumes highlighted the urgent need for more efficient practices. The subsequent market crash in the early 1990s further emphasized the need for a more adaptable and resilient conveyancing sector.

By the mid 1990s the UK conveyancing industry was on its way to becoming more modernized and competitive. Factors such as the Thatcher government's market liberalization policies, the rise of licensed conveyancers and initial technological advances were key in driving this transformation. However, the industry still faced challenges in improving efficiency and adapting to fluctuating market conditions, which would be the focus of further evolution in the ensuing years.

HM Land Registry

In 1940, during World War II, HM Land Registry faced significant operational challenges due to damage from an air raid on central London. To ensure uninterrupted service, the Registry temporarily relocated to the Marsham Court Hotel in Bournemouth, maintaining essential land record services during the war.

The post-war period in the UK saw an increase in property ownership, leading to a heavier workload for the Registry. To manage this, new offices opened in Tunbridge Wells (1955) and Lytham St Annes (1957), helping to distribute the workload and to enhance regional service efficiency. Significant milestones included registering the 1 millionth title in 1950 and the 2 millionth title in 1963, reflecting the Registry's growing importance for national property ownership.

The year 1986 saw a significant transformation, with the Registry's Plymouth office pioneering the production of electronic registers. This shift from paper-based to digital processes significantly improved the efficiency of land registration. The move towards computerization aligned with the goal of implementing comprehensive, compulsory registration across England and Wales.[9]

By 1995 HM Land Registry had not only demonstrated resilience against wartime disruptions but also initiated an essential transition towards computerized operations. This positioned the organization to take full advantage of upcoming technological advances, particularly with the advent of the internet in the property market.

Banks adopt computers

In the late 1950s Barclays Bank set the stage for a revolution in British banking. In August 1959 Barclays made a pioneering decision to order the Emidec 1100 electronic computer at a cost of £125,000. This signalled a groundbreaking shift from traditional manual bookkeeping to electronic computing.

By 1961 Barclays had established Britain's first computer centre, cementing its position as a technological leader in the banking industry. This transition faced initial scepticism from employees and customers, who were both curious and cautious about this new technology. Barclays addressed these concerns by highlighting the benefits of reduced routine work and offering opportunities for staff skill development. This marked the start of a significant shift towards electronic bookkeeping and laid the foundations for the evolution of modern electronic banking, changing financial services forever.

Martins Bank also recognized the competitive edge offered by computing technology, introducing the Pegasus II computer in the late 1950s. This strategic move towards electronic computing was a factor that led to the bank's merger with Barclays in 1968, forming a powerful entity in the banking industry. The merger brought together the technological strengths of both banks, paving the way for remarkable advances in banking operations and influencing the broader UK financial sector.

October 1956 saw a landmark moment in the UK, with the installation of the first non-British computer, an IBM 650, which led to IBM PCs dominating the business computer market over the following decades. In banking, another groundbreaking development occurred on 27 June 1967, with the launch of the world's first cash machine – or ATM – at a Barclays Bank branch in Enfield, London. This innovation allowed customers to withdraw cash using a card and a personal identification number (PIN). The ATM fundamentally changed banking by offering customers a convenient way to access their funds outside traditional banking hours, altering how people managed their money.[10]

In 1959, following the recommendations of the Radcliffe Committee, the National Giro was launched. Operating under the Post Office, the National Giro set itself apart by being the first bank in Europe designed for computerized operations from its inception. In 1984 it pioneered telephone banking: another significant innovation in the industry. This was more than just an additional service channel; it represented a strategic shift towards prioritizing customer convenience and accessibility. This innovation laid the groundwork for modern customer-focused banking services, and it would eventually extend to all business services as we know them today. The National Giro's move into telephone banking was a culmination of decades of technological advances within the UK banking and business sectors, marking the start of an era focused on innovation and customer-centricity.[11]

The 1980s saw further technological advances, notably the rise of client–server computing. This new computing model – in which networked personal computers share resources and information with central servers – revolutionized how organizations managed, accessed and processed data. It optimized various business operations, including transaction processing and customer relationship management.

Several UK institutions and organizations found themselves at the forefront of this era of innovation. The University of Cambridge, in collaboration with Acorn Computers, was instrumental in developing the world's first graphical user interface (GUI) for networking, known as the Cambridge Ring. This breakthrough facilitated efficient information exchange within and between organizations, fostering productivity and growth and cementing the UK's reputation as a centre of technological innovation in banking and business.

Conclusion

The evolution of the property market up to 1996 created a tinder-box foundation that was waiting for the spark of the World Wide Web to ignite the dot-com boom. The interplay of technological advances, legislative changes and economic dynamics during this period primed the industry for the internet's transformative impact.

From the Great Depression through to post-war housing initiatives and into the deregulatory Thatcher years, to the diversification of mortgage products and the shift towards telephone banking and computerized processes, all set the foundation for the introduction of the electronic market that was about to change everything forever.

CHAPTER 3

The arrival of the internet

> The internet is the first thing that humanity has built that humanity doesn't understand, the largest experiment in anarchy that we have ever had.
>
> — Eric Schmidt, former CEO of Google, speaking at the Internet World Trade Show in New York on 18 November 1999

The evolution of the internet – often segmented into the three distinct phases of Web 1.0, Web 2.0 and the emerging Web 3.0 – has fundamentally transformed countless industries, including the property market. This chapter gives a history of the internet's evolution and its effect on the market, starting briefly with the foundational Web 1.0 era for context, but primarily focusing on the pivotal developments of Web 2.0 that led up to the seismic shift caused by the Covid-19 pandemic in 2020. This moment marked the transition from a hybrid online–offline property market to one that was predominantly online and web-enabled.

Over the course of two decades, from the late 1990s to 2019, the property market witnessed the gradual yet significant adoption of web-enabled technologies. This digital transformation was not merely a response to technological advances but was also propelled by a series of political and economic milestones.

The dot-com era

At CERN in 1989 Tim Berners-Lee introduced an idea that would fundamentally change the way we live: the World Wide Web. His vision of a 'universal linked information system' – which used hyperlinks to seamlessly connect a diverse range of information – was groundbreaking. This innovation laid the foundations for the modern internet, revolutionizing how we communicate and conduct commerce globally.

The dot-com era, lasting from the mid 1990s to the early 2000s, was a time of unprecedented growth and speculation in the technology sector, focused especially on internet-based companies. Investors, drawn by the allure of the internet, invested heavily in these companies, despite many not being profitable. The Nasdaq, teeming with tech stocks, reached new highs, creating a speculative bubble marked by inflated stock values and a flurry of initial public offerings (IPOs).

This era was also a time of significant technological advances, with the development of web browsers, search engines and e-commerce platforms. These advances formed the basis for today's digital economy, with companies such as Amazon, Google and eBay either being founded or experiencing major growth during this time.

The late 1990s witnessed a significant milestone in the UK banking sector, with the introduction of the first fully functional online banking platform. It marked the beginning of a new era in financial services – one that embraced the potential of the internet to revolutionize how we manage our finances. Nationwide and Royal Bank of Scotland took the lead when in May and June 1997, respectively, they launched the UK's first internet banking services.[1]

However, by 2000, the bubble had burst, leading to a sharp downturn. The realization that many dot-coms could not sustain their business models or achieve profitability led to massive losses for investors and the collapse of numerous start-ups. The burst of the bubble cooled the tech industry and venture capital investment, and it prompted a period of consolidation for those that survived.[2]

Despite the upheaval, the dot-com era was a pivotal time that transformed the business world, highlighting the internet's disruptive potential and paving the way for the digital economy. It set the stage for the technological innovations and internet services that followed.

From Web 1.0 to Web 2.0

The internet has undergone a radical transformation from its initial incarnation as Web 1.0 to the dynamic and interactive era of Web 2.0. This evolution marks a shift from static web pages to a platform for user-generated content, collaboration and information sharing across various devices.

Web 1.0: the dawn of the digital age

In its early days the internet was a collection of static web pages. This Web 1.0 era was defined by a 'read-only' web, in which content was created by a few and consumed by many. Key players during this period included companies such as Netscape, which provided one of the first web browsers. Desktop-based software was predominant, and the speed and accessibility of online content was limited by dial-up connections. The business model of Web 1.0 was focused on selling software or providing services through a physical medium.

Transitioning to Web 2.0: an interactive web

In October 2004 a pivotal conference hosted by O'Reilly Media marked the christening of a revolutionary concept: Web 2.0. This new phase introduced a dynamic, interactive web, in which users were not just consumers of content but creators as well.

The rise of Web 2.0 coincided with the launch of social media giants such as Facebook in 2004 and Twitter in 2006, which played critical roles in fostering online communities and ecosystems. With its cloud-based services and emphasis on data and user participation, Google epitomizes the Web 2.0 era. Its search engine – driven by algorithms that harness user data to improve search results – and its suite of applications such as Gmail and Google Docs transformed the way people accessed and shared information online. Wikipedia also emerged as a groundbreaking platform to which anyone could contribute, embodying the open-source, collaborative spirit of the time.

This era saw the birth of interactive fan clubs and various online communities, and control over content creation shifted from a central authority to the hands of the users themselves. People could now comment, like, share content or even start their own blogs, connecting with others across the globe in ways previously unimaginable.

Key differences between Web 1.0 and Web 2.0

The transition from Web 1.0 to Web 2.0 marked a significant shift in how the internet was used and experienced. Web 1.0 was characterized by static HTML pages, with users largely being passive consumers of information. Businesses during this era typically sold software and services as products (rather than providing them through 'as a service' models), often via physical stores or mail order.

In contrast, Web 2.0 ushered in a new age of interactivity and user participation. The rise of technologies such as AJAX (asynchronous JavaScript and XML) allowed for dynamic content and responsive user interfaces without the need for constant page reloads. This fostered a two-way communication model, whereby users could actively contribute content, provide feedback and participate in the development of online services.

Web 2.0 also saw a fundamental change in business models. Companies began to use advertising, subscription models and freemium services, generating revenue through online activities and user data. This shift was driven by the realization that data was not just something to be stored and accessed but a core component of the user experience. Personalized services and content tailored to individual user preferences and behaviours became the norm.

This seismic shift in internet culture emphasized the power of connection and community and laid the groundwork for the rich, interactive web experience we enjoy today.

Rightmove

Established in 2000 by industry leaders such as Countrywide, Connells, Halifax, and Royal and Sun Alliance, Rightmove emerged from the dot-com era as a transformative force in the UK property market. Its launch marked the dawn of Web 2.0 businesses for the industry.[3]

Rightmove revolutionized the way properties were bought and sold by providing an online portal that made searches and transactions more accessible. The days of exhausting physical searches were over; comprehensive listings became available with just a click. For estate agents, the platform brought radical changes, reducing their reliance on traditional marketing and significantly expanding their reach.

From its inception, Rightmove has embodied the essence of the Web 2.0 business model, demonstrating the key principles outlined by Tim O'Reilly that define this era of the internet.[4] As an internet-only business, the platform leveraged the full potential of Web 2.0 to transform the property market with its innovative approach.

Rightmove's competitive edge – and most valuable asset – is its unique and comprehensive database of property listings. This rich source of data, difficult for competitors to replicate, makes the platform a go-to destination for both property seekers and property sellers, positioning data as its most valuable asset. By enabling agents to add their listings, Rightmove ensures a constantly updated and expansive property database. Each new listing increases the site's value to other users, and the platform therefore benefits from strong network effects. This interconnected ecosystem of users and data has been instrumental in Rightmove's sustained growth and market dominance.

Embracing the concept of the perpetual beta, Rightmove operates as a 'software as a service' platform, continuously updating and refining its features. This approach allows the platform to adapt swiftly to market changes and user feedback, and to thereby ensure a constantly evolving and improving user experience.

True to its Web 2.0 principles, Rightmove thrives on cooperation, offering web services interfaces and content syndication. This open approach amplifies the platform's utility and reach by fostering a collaborative environment in which data and services are shared.

Understanding the importance of accessibility, Rightmove ensures that its services are not confined to a single device. This foresight has allowed it to stay relevant in an era in which users expect seamless experiences across all their devices, from PCs to smartphones.

Rightmove's emergence as a leader in the Web 2.0 online market can be seen as a result of its excellent execution of the Web 2.0 business-model patterns: prioritizing data, encouraging user engagement, leveraging network effects, committing to ongoing innovation, fostering collaboration and ensuring accessibility across devices. This approach has allowed it not only to lead the market but also to significantly influence the market's direction, establishing a model for success in the online property sector.

A landmark moment in Rightmove's history came with its public debut on the London Stock Exchange in 2006, a move that underscored its innovation and market impact. But its journey did not end there, and it has continued to evolve and adapt, particularly in the face of emerging competition. The launch of

Zoopla in 2008 and OnTheMarket in 2015 brought new challengers to the online property portal scene, with each introducing fresh technological innovations and strategies aimed at capturing market share.

Despite the competition, Rightmove has successfully maintained its market-leading position. The platform's commitment to continuous improvement and technological integration has been pivotal in enhancing user experience and meeting the evolving demands of both consumers and the industry at large. By staying ahead of technological trends and prioritizing the needs of its users, Rightmove has not just survived the competitive pressures but thrived, reinforcing its status as a benchmark for excellence in the digital property market.

Estate agents and panel companies

From the advent of the internet up to 2019, the UK estate agency market witnessed a transformative shift, characterized not only by online integration and strategic partnerships but also by the emergence of online brands and key mergers and acquisitions.

Brands such as Purplebricks and Yopa, with their lower fees and online-centric services, challenged traditional models and offered property sellers alternatives to high street estate agents. These online agencies capitalized on technology to streamline the selling process, providing platforms for listing, viewing and managing property sales with unprecedented efficiency.[5]

The period also saw significant consolidation within the industry. Notably, the merger of Countrywide and Bairstow Eves created one of the UK's largest property services groups. Another significant move was the acquisition of Hamptons International by Countrywide, further expanding its portfolio of property services.

Traditional estate agents responded to the digital challenge by enhancing their online presence and services. Foxtons, Savills and Knight Frank invested heavily in digital marketing, online property listings and customer engagement tools in order to compete in the changing market. They developed sophisticated websites and mobile apps, offering virtual tours and leveraging big data for targeted marketing campaigns.

As referral services became established as a large and profitable market, companies were set up to, in effect, take leads from estate agents and sell them to conveyancers and brokers. These marketing companies became known as panel companies, as they would collate a group of mortgage brokers or conveyancers, pay estate agents for leads and then sell these leads to the conveyancers or brokers. Over time, these companies sought to add additional value by curating their panel members, and in some cases by ensuring the quality of members' services. They also began providing agents or banks with progress updates on the property transactions and other work that had been 'panelled out'. In recent years, the referral fee business model has become a source of national concern, which has led to reviews by Trading Standards and by the most recent government inquiry into the property market.[6]

The 2007–8 financial crisis (discussed in more detail below) was another pivotal moment, prompting agents to seek more cost-effective marketing channels and to diversify their revenue streams. The crisis accelerated the adoption of online marketing strategies, the fostering of social media engagement and the use of analytics for market and customer insights.

The distinction between traditional and online agents now began to blur as hybrid models emerged. Offering the convenience of online transactions with the personal touch of traditional services, these models aimed to combine the best of both worlds, and to provide flexibility and choice to consumers.

Towards the late 2010s there was a growing emphasis on sustainability within the property market. In response to increasing consumer demand for environmentally friendly homes, estate agents began to promote properties with green features, such as energy efficiency and sustainable materials.

The financial crisis

Though its causes and effects extended well beyond the evolution of the internet, the financial crisis of 2007–8 is crucial for understanding the period from which the online property market emerged. Its profound impact on the UK property and mortgage markets was largely due to the collapse of key financial institutions and the subsequent liquidity crisis. Northern Rock's aggressive expansion, fuelled by international money markets, left it exposed when these sources dried up as a result of the US subprime mortgage crisis. This led to the first UK bank run in more than 140 years, shaking confidence in the banking sector and triggering a government guarantee of deposits to prevent further panic.[7]

The crisis highlighted the vulnerability of the UK's housing finance system, which had grown alongside the significant expansion in home ownership. The government's intervention to stabilize Northern Rock and other financial institutions underscored the property market's interconnectedness with banking stability.

As credit became scarcer and more expensive during the global crisis, the UK's mortgage market was directly impacted. Mortgage approvals plummeted, and the cost of borrowing increased for households, contributing to a sharp decline in property transactions and falling house prices. The liquidity crisis meant fewer mortgage products were available to consumers, and the previous decade's rapid growth in home ownership halted as a result.

The crisis also led to significant changes in the regulation and oversight of the mortgage market in the UK. Banks became more cautious in their lending practices, requiring higher deposits and stricter affordability checks for mortgage applicants. This shift made it more difficult for first-time buyers to enter the market and further slowed the rate of home ownership growth.

Moreover, the financial instability brought about by the crisis resulted in the merger of Lloyds TSB and HBOS, which restructured the banking sector and reduced competition in the mortgage market. This merger, along with the nationalization of Bradford & Bingley and the sale of its assets to Santander, was part of a broader reconfiguration that impacted mortgage lending policies and practices in the UK banking industry.

The government's response to the crisis, including the extension of the Special Liquidity Scheme by the Bank of England, was pivotal in preventing a complete collapse of the mortgage market. However, it also marked the beginning of a more cautious era in UK property finance, with long-lasting effects for affordability, lending standards and the dream of home ownership for many Britons.

E-government

The late 1990s brought the UK government's push towards electronic government, or e-government, with the ambition of making public services fully accessible online. This initiative aimed to transform the internal efficiency of government departments through technology, and to thereby enhance operations and public access.[8]

HM Land Registry stood out as a leader in e-government efforts. It achieved significant operational efficiencies, notably in its reduction of physical document handling. This not only sped up processes and reduced errors but also streamlined operations, allowing the Land Registry to cut down on unnecessary steps and manage more work without increasing staff. A striking achievement was the reduction of post-completion application processing times from three weeks in the mid 1990s to just two days by 2000.

Another landmark achievement was the June 2000 launch of Land Registry Direct, which offered online access to registers, plans and services. This greatly impacted usage: the number of account holders grew to more than 3,700, and monthly register views jumped from about 45,851 in January 1999 to 186,196 in August 2001. Electronic notifications of discharge also saw a significant increase, from 6,000 in April 2000 to 85,885 in August 2001.

The Land Registry's digital shift was strategic and customer-focused, enhancing service delivery and operational efficiency. By adopting digital solutions, the Land Registry not only adapted to technological changes but also set a benchmark for digital accessibility and efficiency in public services.

In July 2001 a six-year collaboration between the Law Commission and the Land Registry culminated in a groundbreaking report: 'Land registration for the twenty-first century: a conveyancing revolution'. This report marked a pivotal moment in land registration by advocating for the dematerialization of conveyancing processes. Dematerialization would mean moving away from the traditional approach – in which a buyer's title is deduced from documents dating back more than 15 years – to a system in which the title is confirmed solely by the register: 'It will be the fact of registration and registration alone that confers title.'[9]

The enactment of the Land Registration Act 2002 on 26 February 2002 provided the legal framework necessary for implementing e-conveyancing, setting a new direction for property transactions in the UK.

Building the property market infrastructure

During this period there was a recognition that the property market was missing an infrastructure. There were two notable attempts to supply it.

Land Registry Network

It was envisaged that the Land Registry Network would use internet technology to move the property market from a paper-based system to one that was fully electronic. This new online network would solve several issues with the paper-based system. One prominent issue is 'the registration gap', which arises from the fact that the legal title of the land registered to a property does not pass to the buyer until the transfer or the new lease is registered at the Land Registry. The registration gap is therefore the period between the completion of the purchase (when the funds are transferred and the keys pass to the buyer) and the completion of the application for registration at the Land Registry (when the updated title is created, noting the buyer as the registered proprietor).[10]

Additionally, the proposed system was designed to manage transaction chains effectively. The secure electronic communications network at the heart of this system would monitor chains of transactions in order to identify delays and facilitate prompt remedial action. The expectation was that this would lead to expedited transactions and fewer broken chains, significantly improving the efficiency of property dealings.

In 2006 the Land Registry embarked on realizing the core capabilities of the Land Registry Network. The project of designing and implementing the network was called Chain Matrix and was led by IBM. The Land Registry's 2006 annual report mentioned that the project was getting ready for its initial launch:

> We will be launching the first tranche of e-conveyancing services which, notably, includes a prototype chain matrix service. The chain matrix allows all the transactions in a residential conveyancing chain to be viewed online, identifying bottlenecks and providing transparency. It will be a key feature of our full service and the prototype is designed to give us valuable experience in creating this final version. Electronic conveyancing represents our biggest investment and our greatest risk.[11]

Sadly, the project was not successful, and the annual report a year later communicated that it was being paused:

The Land Registry Board decided in December 2007 to focus resources on electronic discharges [and defer work on Chain Matrix]. The decision was influenced by valuable feedback from our customers following a prototype of Chain Matrix, which ran for six months. Land Registry is still committed to Chain Matrix and EFT [electronic funds transfer], but evolving technology means there is no guarantee that we will benefit from the development work already undertaken.[12]

The project was subsequently abandoned and was not mentioned in the following year's annual report.

Veyo e-conveyancing platform

Seven years after the shelving of the Land Registry project, the Law Society launched Veyo, another effort to establish an e-conveyancing platform. This initiative led to a partnership with Mastek, an Indian outsourcing firm, and the creation of the joint venture vehicle Legal Practice Technologies Limited. The Law Society had a 60% stake and Mastek the remainder.

However, within a year it became apparent that this project was also facing significant hurdles. A new chief executive was appointed to lead Legal Practice Technologies, and shortly thereafter a decision was made to dissolve the company by December 2015, culminating in an £11 million loss.

Right objective, wrong delivery model

One common aspect of both Chain Matrix and Veyo is that they were outsourced, one to IBM and the other to an Indian outsourcer. Having spent my entire career in technology and having dealt extensively with outsourcing, I find the clear takeaway to be that while neither initiative was a bad idea in itself, the outsourcing model – with its commercial rigidity and waterfall delivery approach – is ill-equipped to deal with innovation projects.

It was a genuine setback for the property market when the Chain Matrix and Veyo initiatives failed to deliver a common, national market infrastructure. The conceptual foundations laid out in the Law Commission's report, the legal framework created by the Land Registration Act and the documentation produced by the Chain Matrix project precisely capture the kind of infrastructure that the market needed then and still needs today. Chapter 5 will cover this missing infrastructure in detail.

Conclusion

This twenty-year period saw an extensive transformation in the UK's property market, largely driven by the advent and evolution of the internet and associated digital technology. The journey began when Tim Berners-Lee's invention of the World Wide Web ignited the dot-com era, leading to a boom in internet-based companies and speculative investment. Despite the bubble's eventual burst, this early phase set the stage for the rise of e-commerce and digital communication.

The launch of Rightmove in 2000 marked a pivotal shift towards online property listings, revolutionizing how properties were bought and sold. This era also saw the emergence of online estate agencies such as Purplebricks and Yopa, which challenged the traditional estate agent model by offering lower fees and online services. The rise of panel companies and comparison sites led to yet another change in how consumers and industry professionals interact with property and mortgage services.

The late 1990s had introduced the UK to online banking, further digitalizing financial transactions and impacting the property and mortgage markets. However, the financial crisis of 2007–8 imposed a more cautious approach on property finance, with tighter regulation and oversight of the mortgage market.

The government's push towards e-government and the digital transformation of the Land Registry aimed to modernize conveyancing processes, though these efforts have not been without their difficulties, as seen in the setbacks faced by the Chain Matrix project and the Veyo e-conveyancing platform. Digital advances, regulatory changes and macroeconomic challenges therefore all played a part in reshaping the UK property market into an online market. But as we will see, this online market has failed to deliver on the promise of the internet and in many respects is broken.

CHAPTER 4

The online market is broken

> There are more than 1,000,000 residential and around 90,000 commercial property transactions in England and Wales each year. Despite advances in digital and online technology, there is evidence that the average 2021 home sale took 49% longer to complete than in 2007. Research shows that more than a quarter of all property transactions in England and Wales do not complete, at an estimated total financial loss of around £400 million for those trying to sell.
>
> — HM Land Registry Strategy 2022+[1]

In this chapter I aim to briefly highlight the principal challenges currently facing the online property market, with subsequent chapters going into more detail.

In today's market, consumer interactions are highly fragmented. This is a cause for concern among all stakeholders, including buyers, sellers, property professionals and regulators. There is a general consensus that the market is not functioning optimally for any of these groups.

Two comprehensive reports shed light on these challenges. The first is the report on 'Improving the home buying and selling process' published by the Ministry of Housing, Communities and Local Government in 2018.[2] The second, containing valuable reference material, is a detailed review titled 'Improving the home buying and selling process in England', a House of Commons research briefing published in 2022.[3] These reports provide a thorough analysis of the current state of the property market and offer insights into potential solutions.

An earlier white paper with more detail on the issues from the perspective of conveyancers was published in 2016 by the Conveyancing Association. The wide-ranging document, titled 'Modernising the home moving process', examines the systemic problems that have led to widespread dissatisfaction among all parties involved in the conveyancing process.[4]

The association's findings pinpoint three primary areas of concern that have significantly affected the efficiency and reliability of today's online market: the lack of transparency, the lack of certainty and the prevalence of delays. The report elaborates on how consumers often find themselves navigating this complex process without sufficient information or an adequate understanding of their options, which contributes to the lack of transparency. Furthermore, the absence of binding offers and the unpredictability of move dates exacerbate the lack of certainty, while a series of identified bottlenecks within the transaction process leads to frustrating delays.

Drawing on insights from other jurisdictions where the home moving process is more streamlined, the Conveyancing Association suggests several reforms. Among these, an early legal commitment to transactions stands out as a measure that could significantly reduce delays and eliminate the domino effect of transaction chains. Additionally, the report advocates the upfront provision of property information to all parties involved, echoing successful practices that have led to quicker, more transparent transactions in other countries.

As I was completing this book, in March 2024, the Levelling Up, Housing and Communities Committee launched an inquiry on improving the home buying and selling process. Clive Betts, chair of the committee, gave this summary of the problem:

> The process of buying and selling a home in England is often stressful for those involved. Indeed, despite there being around two million households who successfully buy or sell

their home each year, consumers often find the process is not as efficient, effective, or as consumer-friendly as it could be.[5]

The inquiry intends to investigate the chief obstacles to improving the process of buying and selling a home, and in particular the time taken to complete a transaction, the difficulty of finding the right information, the lack of transparency around conveyancing services, the payment of referral fees and the weak regulation of estate agents

Every party involved in property transactions experiences pain; none are well served by the current market structure. For consumers especially, buying and selling property online is a challenging, frustrating and stressful experience, and very different to the experience of any other online service they are used to – planning a holiday, for example, can be effortlessly managed through platforms such as Expedia.

From a consumer's perspective, then, a key challenge lies in the absence of a single platform for seamlessly completing end-to-end property transactions. Of course, the businesses that serve consumers – such as agents, brokers, conveyancers and banks – all aim to have the consumer use their platform and to 'own' the customer relationship directly. This desire, understandable as it is, leads to a situation in which consumers are required to create accounts and provide the same data across a plethora of disconnected platforms and portals.

This segmented approach forces consumers to coordinate with multiple professionals representing these different businesses, each operating autonomously. Such a scattered system demands a significant effort from consumers to collate and synchronize information across disparate platforms. This not only adds complexity to the transaction process but also raises serious data security and privacy concerns.

The financial ramifications of the resulting transaction breakdowns are significant and far-reaching. Over half of the buyers in these cases find themselves incurring expenses for services such as property surveys, valuations, mortgage arrangements and legal fees, even though the transaction did not reach fruition. On average, the cost burden for a buyer in a transaction that falls through ranges from £695 to £744. Sellers, too, are not spared from this financial strain, with their costs in such scenarios typically falling between £582 and £740. Alarmingly, about a quarter of both parties – buyers and sellers – face expenses that exceed the £1,000 mark.[6]

For estate agents, winning instructions from sellers to act for them and list their property is crucial, as their revenue typically materializes only on transaction completion. This structure necessitates stringent cash flow management, with any delay in completion directly impacting their financial health. A key performance indicator for estate agents is their pipeline turnover rate, which measures the efficiency of converting potential sales into completed transactions. But due to the ever-increasing complexity in the online market, agents are reporting a continual reduction in these rates. Ironically, the same technology intended to streamline processes often contributes to this complexity and adds to delays, as it is stand-alone and not interoperable with internal and partner systems.

Conveyancing firms rely on a steady flow of clients, given the typically non-recurring nature of property transactions. Once they secure a client, the focus shifts to processing the transaction efficiently. However, a lack of streamlined digital connectivity among relevant parties means that conveyancers often have to act as project managers for the transaction, resulting in a host of administrative and coordinating tasks. Taking on more and more of these activities inflates operational costs for conveyancers and restricts the number of cases each professional can manage, ultimately affecting the firm's profitability and operational efficacy.

Mortgage brokers, acting as advisors to buyers, help steer the mortgage process from application to completion. The market's disjointed nature, however, leaves them with limited visibility of the progress of transactions, hampering their ability to provide clients with timely updates. Like estate agents, brokers are adversely affected by the high rate of transaction failures, which can significantly impact their income.

Lenders also face challenges in delivering a seamless mortgage journey. While the initial stages of sourcing and applying for a product are often digital, the process fragments post-application. The necessity of interacting with conveyancers and surveyors via disparate systems introduces inefficiencies and elevates operational costs and risks. This lack of a unified data infrastructure not only impedes a smooth digital experience for customers but also increases operational risks and costs for mortgage lenders.

Surveyors operate in a challenging environment marked by a mix of manual and electronic processes. The absence of comprehensive property information at the outset can lead to post-valuation queries and consequent delays. Moreover, communication breakdowns with homeowners, lenders or agents, exacerbated by the lack of a common communication platform, can further hinder their operations.

HM Land Registry, positioned at the culmination of the property transaction process, itself often faces significant challenges due to fragmented data flows within the industry. This fragmentation frequently leads to the submission of poor-quality and inaccurate data to the Land Registry, which in turn results in a high volume of post-completion queries. As well as creating additional work and associated costs for the Land Registry and conveyancers, these issues also expose both lenders and consumers to considerable risks arising from the 'registration gap'.[7]

These data quality issues have multiple negative impacts. For the Land Registry, this means dedicating substantial resources to addressing and rectifying inaccuracies. This not only leads to increased operational costs but also slows down the overall process of updating and maintaining accurate property records. A Land Registry blog post describes the scale of the challenge:

Every day, the Land Registry receives around 18,000 applications to update the register. The majority of these applications are submitted correctly with all the necessary information included for our land registration experts to complete the requested change. However, for one in five applications – more than 3,500 applications every day – some of the information needed is missing, incomplete or wrongly drawn.[8]

The current challenges faced by the Land Registry highlight a critical need for improved data management and communication practices within the property market.

Missing market infrastructure

The property market in the UK currently faces a significant systemic challenge due to the absence of a cohesive market infrastructure. Unlike other industries – particularly financial services, where financial market infrastructures play a pivotal role in streamlining processes and ensuring efficient transactions – the property sector operates without basic data foundations. This gap has profound implications for the efficiency and effectiveness of property transactions.

In the property market, the lack of such an infrastructure means that key stakeholders – including buyers, sellers, estate agents, conveyancers, mortgage lenders, surveyors and the Land Registry – work in digital silos. As a result the market runs on emails, telephone calls and a whole host of disconnected systems. Each party relies on its own protocols, and this fragmentation leads to several inefficiencies in property transactions: duplicated efforts in data collection and verification, delays due to a lack of coordination and increased risks of errors.

What are the key challenges? They are the absence of a secure digital identity, the prevalence of fragmented data and the lack of a seamless settlement system.

The absence of a unified, shareable identity system underpinned by a comprehensive identity trust framework means that there is no standardized digital identity across the property transaction process, leading to a fragmented and inefficient ecosystem that affects everyone involved. A direct consequence of this fragmentation is the necessity for consumers to undergo multiple identity verifications at different stages of a property transaction. This process is not only inconvenient but also poses significant data privacy and security risks. Consumers are required to repeatedly share their sensitive personal information with various entities. Each instance of sharing increases the risk of data breaches and misuse, exacerbating consumer apprehensions about digital safety.

The issue of significant data fragmentation stems in large part from the lack of standardized data formats across the industry, leading to inefficiencies and heightened operational costs. The main difficulties are around finding necessary data (such as upfront property information), accessing it even when its location is known, exchanging data between systems in the absence of a standard, and re-using data that has often lost its context and provenance. Consequently, errors and delays in processing transactions become more frequent, resulting in increased chances of transaction failures. Moreover, the absence of a trust framework for secure data sharing makes the property transaction process slower and riskier, with a higher chance of data breaches and the loss of sensitive information.

The completion stage of transactions relies on outdated, inefficient practices that present challenges to all stakeholders. Despite advances in digital technology, the completion process still requires manual operations across disjointed electronic systems. This discrepancy creates significant delays and uncertainties, especially with the manual handling of funds and the updating of property titles at the Land Registry. The slow, cumbersome escrow systems and the frequent registration gaps introduce risks and legal ambiguities that compromise the security and efficiency of transactions.

Conclusion

This outdated approach affects everyone involved in property transactions.

Consumers, whether buyers or sellers, are facing frustration due to delays and a lack of trust in the process. Their journey through the property market is anything but straightforward. They must navigate a maze of platforms and services without a central guide, which results in concerns over data privacy and a fragmented experience. This disconnection often ends in transaction failures, adding financial stress to an already tense process.

On the professional side, the story is not much different. Estate agents, conveyancers, mortgage brokers, surveyors and lenders each face their own set of hurdles due to the market's lack of data infrastructure. Estate agents are trying to keep their pipelines moving and maintain financial health. Conveyancers and mortgage brokers are drowning in paperwork and struggling for a clear view of their transactions. Legal professionals and conveyancers face a hefty administrative workload, reducing their efficiency and increasing costs. Financial institutions and lenders also contend with the repercussions of these inefficiencies, as the slow movement of funds and registration gaps complicate lending processes.

Another key problem lies in how transactions are completed. The current manual and fragmented approach not only slows things down but introduces risks and inefficiencies.

Beyond these immediate challenges are deeper, systemic issues. The property market lacks a solid foundation, with no unified market infrastructure or standardized digital identity. The concept of 'smart data' – findable, accessible, interoperable and reusable – is still far from being a reality. These missing pieces lead to operational inefficiencies, data problems and elevated risks, all of which dampen the market's overall performance. The market desperately needs a digital transformation. A unified market infrastructure, standardized digital identities, the adoption of smart data principles and a modern approach to completing transactions are not just nice-to-haves – they are necessities. These changes are vital for boosting the market's efficiency, reliability and security, for restoring consumer confidence, and for ensuring the property market can meet the demands of a rapidly evolving digital world.

PART II

DIGITAL FOUNDATIONS

To fix the current online property market and achieve a situation in which consumers and the businesses that serve them can participate in cohesive property transactions, the market needs to invest in some basic digital foundations. This work has already commenced, and in my role as COO of Coadjute I have been involved in many of these initiatives. In the following chapters I provide an overview of the key market activities that are focused on building robust digital foundations for the market.

CHAPTER 5

Market infrastructure

> [The Land Registry Network] enables the registrar to establish an electronic communications network, either himself or through a third party, which will be used as he sees fit in connection with registration, and with the carrying out of transactions which involve registration and are capable of being effected electronically.
>
> — Land Registration Act 2002[1]

In this chapter I look into the consequences of the missing property market infrastructure and contrast this situation with that in the financial sector, where financial market infrastructures (FMIs) such as Visa, Mastercard and Swift have long been in place.

As I emphasize throughout the book, the need for a market infrastructure is neither new nor unknown. The Land Registration Act 2002 detailed the creation of a property market infrastructure known as the Land Registry Network. However, as previously discussed, the network was abandoned.

Drawing inspiration from FMIs in the financial sector, I propose that we utilize these models to revolutionize property transactions. Contrary to the belief in some quarters that application programming interfaces (APIs) alone can resolve the issue of connecting various systems and services in the property market, I advocate for a fully developed property market infrastructure that guarantees standardized, secure and seamless data integration between different systems.

To illustrate this point, I present my own company, Coadjute, as an example of a trusted data ecosystem for businesses within the market. This case study demonstrates both the feasibility and the necessity of applying FMI principles to the property market. Concluding the chapter, I urge for a collaborative effort between the industry and government to establish a digital backbone that enhances the efficiency and security of the property market.

The market is missing an infrastructure

Our world is composed of layers, with infrastructure at the foundation of it all. Infrastructure – be it physical, like roads and railways, or digital, like communication tools such as email – plays a pivotal role in enabling markets to function. The absence of a property market infrastructure not only complicates property transactions but also adds unnecessary layers of inefficiency and risk.

UK mortgage industry statistics for the end of 2022 show that the value of all residential mortgage loans was £1,675 billion, almost 4% more than the final figure for the previous year. The value of gross mortgage advances in 2022 Q4 was £81.6 billion.[2] It is hard to believe that despite more than £350 billion in property assets changing hands every year – a value exchange that is material to the UK economy – there is still no market infrastructure. Furthermore, there is not sufficient recognition that this is a systematically risky and economically inhibiting situation for the UK.

The absence of infrastructure for the property market contrasts with the state of financial services; it is easy to transfer money, but not a property title. This discrepancy often invokes a reflexive response: a house is a high-value asset requiring stringent governance and control. True as this may be, it need not entail a convoluted and inefficient process.

Certainly, there should be some distinctions in handling these transactions. Cash, being a fungible asset, is interchangeable with any other equivalent amount. In contrast, a house, as a non-fungible asset, is unique. Verifying the value of £285,000 in cash primarily involves confirming the legitimacy of its source. Establishing the value and ownership of a £285,000 property asset is more complex. Yet, setting aside the tasks of verifying valuation and ownership, it is reasonable to expect that the transfer process for a property could be as streamlined as it is for money. Ultimately, in both instances, what we aim to do is update one or more ledgers or registries – essentially, databases.

This comparison between transferring property and money is often overlooked. Underpinning money transfers is the network of systems known as FMIs. These are often described as the 'plumbing of the financial system' – an image that underscores their role as the fundamental networks facilitating financial dealings. They encompass the platforms and mechanisms through which transactions are processed, cleared and recorded, enabling interactions between individuals, institutions, businesses and other market participants.

The concept of FMIs plays a critical role in understanding the challenges and potential solutions for the UK's electronic property market.

Learning from financial market infrastructures

It is encouraging to read that, back when HM Land Registry was thinking about how to design and implement a property market infrastructure, it looked at what had been achieved in financial services: 'Following the precedent that was adopted to introduce electronic trading in securities, we consider that the new system of electronic conveyancing should be introduced by rules made under a wide statutory enabling power.'[3]

There is much that can be learned from financial market infrastructures, including three of the biggest: Visa, Mastercard and Swift.

The origins of credit cards

In 1950 Frank McNamara, a New York businessman, found himself in an embarrassing situation while dining at a restaurant when he realized he had forgotten his wallet. This incident sparked an idea: a charge card that could be used at multiple establishments, eliminating the need to carry cash or individual credit cards for each merchant.

McNamara partnered with his friend Ralph Schneider to establish Diners Club, the world's first independent credit card company. The Diners Club card, which debuted in 1951, was initially made of paper and had to be paid in full each month. It was accepted at a limited number of restaurants and hotels, primarily in New York.

Despite its limited scope, the Diners Club card gained popularity among businessmen and travellers, who appreciated the convenience and flexibility it offered. In 1952 Diners Club introduced the first plastic credit card, which became the standard for the industry.

As Diners Club grew, it expanded its services internationally, becoming the first credit card company to be accepted worldwide in 1953. This global presence demonstrated the potential for credit cards to facilitate cross-border transactions and paved the way for future competitors such as Visa and Mastercard.

Visa

Visa has a history that is closely intertwined with the development of the credit card industry and the broader narrative of modern financial services. Marked by innovation, strategic partnerships and a keen focus on technological advance, its origins and evolution reflect significant milestones in the journey towards a cashless global economy.

Visa's roots can be traced back to 1958, when Bank of America (BofA) launched the BankAmericard programme in Fresno,

California. This initiative represented the first mass mailing of unsolicited credit cards, a bold experiment aimed at providing consumers with a revolving credit line that could be used at a variety of merchants. The programme's initial rollout was fraught with challenges, including high rates of delinquency and fraud, but it eventually laid the groundwork for the modern credit card system.

Recognizing the potential for a national credit card system, BofA began licensing the BankAmericard programme to banks outside California in 1966, under the condition that they adhere to uniform standards and interchange fees. In 1970 the various BankAmericard licensee banks formed National BankAmericard Inc. (NBI), an independent corporation that would oversee the administration, standardization and promotion of the BankAmericard system. This was a pivotal step towards creating a cohesive, interoperable network of banks and merchants.

In 1976 NBI underwent a significant rebranding and organizational restructuring, adopting the name 'Visa' to reflect its growing international presence and the vision of universal acceptance. The name was chosen for its simplicity, ease of pronunciation in multiple languages and connotations of universal access.

Visa's organizational structure has been characterized by its emphasis on collaboration among member banks, on technological innovation and on creating a secure and efficient global payment network. Visa operates on a membership model that sees banks issue Visa-branded cards and manage customer accounts while Visa itself focuses on processing transactions and ensuring network security.

Visa's governance structure has historically included a board of directors composed of representatives from member banks. This structure ensures that the interests of the banks that issue Visa cards and process transactions are closely aligned with Visa's strategic direction. Visa has been at the forefront of payment technology, introducing innovations such as electronic authorization of transactions, security protocols such as chip-and-PIN, and more recently, contactless payments and digital wallet services.

Mastercard

In the mid 1960s, seeing the success of these early credit cards, a group of banks sought to create a competitor that would come closer to being universally accepted and not be tied to a specific bank or company. In 1966 this led to the formation of the Interbank Card Association (ICA), which would later become Mastercard. The ICA was a collective of several banks that aimed to create a credit card system that was accessible to a wider range of banks and merchants.

The ICA introduced its credit card, initially called 'Master Charge: The Interbank Card', in the same year it was founded. This was a significant step towards creating a universal credit card that could be used beyond local or national boundaries. The name 'Master Charge' was chosen to signify the card's wide acceptance and the financial backing of the member banks.

From its inception, Mastercard's organizational structure was designed to be cooperative, with ownership and governance shared among the member banks. This structure was crucial for gaining widespread acceptance among both financial institutions and merchants.

As with Visa, the governance of Mastercard has traditionally been overseen by a board of directors composed of executives from member banks. This board decides on the strategic direction of the company and the policies and fees associated with card usage.

Initially, member banks were primarily from the United States, but the organization quickly expanded internationally. Member banks issued cards under the Master Charge (and, later, Mastercard) brand, handled customer accounts and managed merchant relationships. The operational model involved the central processing of transactions, which allowed for efficiency and security in handling credit card payments across different banks and merchants worldwide.

The rebrand from Master Charge to Mastercard came in 1979, as part of a strategy to emphasize the card's versatility beyond mere 'charges' and to reflect its growing international presence. Throughout the 1980s and 1990s Mastercard continued to innovate, introducing magnetic strips for easier transaction processing and later adopting chip technology for enhanced security. The company also expanded its global footprint, establishing itself as one of the leading payment networks worldwide.

In 2006 Mastercard underwent an IPO and transitioned to a publicly traded company. This move allowed Mastercard to invest more heavily in technology and expansion, although its governance still reflects the interests of its founding members – the banks.

Swift

In the financial world, Swift (the Society for Worldwide Interbank Financial Telecommunication) keeps money moving smoothly across the globe.[4] Founded in 1973 by a group of 239 banks from fifteen countries, Swift was born out of a need for a more efficient and secure way for banks to send each other messages about money transfers and other transactions. Before Swift, banks relied on the Telex system, which was slow, not standardized and often insecure.

Swift is universally recognized as the world's most secure and reliable third-party network, catering to 212 countries and supporting over 10,000 banking organizations, securities institutions and corporate customers

The network revolutionized banking communication by offering a standardized, reliable system for financial institutions worldwide. Unlike Visa or Mastercard, which deal directly with consumers, Swift operates behind the scenes, ensuring that banks can talk to each other securely and efficiently.

Owned by its member banks, Swift is a cooperative that functions under Belgian law, with oversight from the National Bank of Belgium. Its democratic structure means that every member bank has a say in its operations, from setting policies to electing the board of directors who steer the ship. The day-to-day running of Swift is in the hands of its executive committee, led by its CEO.

As the financial world evolves, so does Swift. The organization is not just about sending messages anymore. It has been at the forefront of banking innovation, introducing new standards and technologies to make international transactions faster, more transparent and more traceable. Its latest venture, Global Payments Innovation, is a testament to its commitment to improving the speed and reliability of cross-border payments.

Land Registry Network: defining market infrastructure

The Land Registration Act 2002 laid the groundwork for introducing a property market infrastructure (PMI) known as the Land Registry Network, which aimed to facilitate the electronic execution of property transactions that require registration.

The notes to the Act stated:

> The Act creates a framework in which it will be possible to transfer and create interests in registered land by electronic means. It does so by enabling the formal documents to be executed electronically and providing for a secure electronic communications network. Because it is envisaged that the execution of those documents and their registration will be simultaneous, and the process of registration will be initiated by conveyancers, permitting access to the network is to be controlled by the Land Registry, which will also exercise control over the changes which can be made to the register.[5]

The Act described a system that would control varying levels of access to the network through access agreements. These agreements, both contractual and enforceable, define the terms and conditions under which various parties can use the network. The registrar is vested with the authority to determine these terms, which may include provisions for charging network-usage fees. This flexible approach is key to accommodating a diverse range of users while upholding the network's integrity and security standards.

A transformative aspect of the Land Registry Network, as envisaged by the Act, is the registrar's ability to mandate the execution of certain registrable transactions electronically. This authority signifies a deliberate move towards phasing out paper-based processes in favour of electronic methods in specified instances, and it underscores the government's commitment to advancing electronic conveyancing and streamlining property transaction processes.

Empowering the registrar or an appointed authority to manage and monitor transactions within the network is especially important for chain transactions, which the Act recognizes as a common source of complexities in property dealings. This capability is aimed at reducing the challenges and uncertainties inherent in chain transactions, enhancing the overall efficiency and predictability of the conveyancing process.

In essence, the Land Registration Act 2002 provided the legislative foundation for developing what we would today call a PMI. It also outlined the functional requirements of such a system. Although the PMI has not yet materialized, the Land Registry's strategy document for 2022 onwards clearly describes a modern version of the network defined in the 2002 Act: 'Our vision is one where connected digital platforms and services enable buyers and sellers, their banks, lawyers and others to join together each time property is bought and sold, interacting digitally and seamlessly.'[16]

To deliver this vision, a PMI or network would require several key components. Integration as a service, for example, would have to be provided to all digital platforms in the market, so that users could connect and interact in managing their shared property transactions. This integration, in turn, would rely on the development and oversight of APIs for effective system communication and data interoperability across different platforms.

Another facet is access control, which would address the challenges around multi-stakeholder access with secure data sharing mechanisms and role-based controls in order to ensure data security and privacy. Real-time synchronization protocols, conflict-resolution mechanisms and comprehensive audit trails would have to be in place for seamless data synchronization across all parties involved. And all these parties would need to have robust digital identities, including a system for digital identity verification and secure authentication protocols.

Data standardization and management is also critical and would involve standardizing data formats, ensuring data validation and quality control and establishing rules for data prioritization in order to resolve conflicts. Real-time data availability and transparency would be a prerequisite for easily tracking transaction statuses and receiving instantaneous updates. Compliance and security measures for this data ecosystem – adhering to legal and regulatory requirements, with robust encryption and data protection – would be non-negotiable.

Finally, establishing mechanisms for user feedback and adopting an agile development approach for iterative enhancements would be crucial for continuous improvement and meeting changing market needs.

The evolution from a simple vision statement to a detailed set of requirements for a PMI reveals the complexity and depth involved in creating a platform that delivers the Land Registry's vision. Creating a PMI that truly transforms the property market requires a comprehensive, holistic approach that addresses not only the technological aspects but also the legal, organizational and operational dimensions. It involves establishing a robust framework that facilitates seamless integration, ensures data security and privacy, and promotes transparency and efficiency throughout the property transaction process.

APIs are not market infrastructure

In the quest for a more connected and efficient property market, the topic of APIs often arises as a potential solution. Many discussions about property market connectivity and system interoperability quickly turn to APIs, with claims that integration between various services in itself constitutes market infrastructure or can deliver market-wide connectivity. However, this notion is a misconception that needs to be addressed.

To illustrate the limitations of relying solely on APIs, let us consider a scenario involving an estate agent and a conveyancing firm, each operating its own digital platform.

The first challenge arises when the estate agent's system attempts to find and connect to the conveyancer's system. With thousands of conveyancers in the market, how does the agent's system 'find' the conveyancer's? And assuming it can, how does it then connect and authenticate? How can the conveyancer's system trust the agent's system? Or, more accurately, how can it have the ability to trust any arbitrary agent systems that request a connection?

Even assuming that access control is performed successfully, the next hurdle is achieving two-way data exchange without a common data standard. Ensuring interoperability, data transformation and reusability between systems is a complex task that goes beyond the capabilities of APIs alone.

Furthermore, consider a situation in which the conveyancer updates critical information, such as a completion date. How does the estate agent's system process and recognize this updated information automatically? Is the data synchronized seamlessly across both platforms? These questions highlight the need for a more comprehensive solution that extends beyond the scope of APIs.

As we transition from a bilateral transaction to a multi-party scenario, the complexity multiplies exponentially. Ensuring consistent data standards, access control and synchronization across multiple parties becomes an even greater challenge. Each additional connection introduces concerns about data consistency and reliability – concerns that again emphasize the need for a robust, market-level infrastructure.

APIs, while facilitating standardized data interaction, do not inherently solve these challenges. APIs lack an understanding of business contexts, business logic, workflows and the state of the data, all of which require a sophisticated, intelligently designed market-level infrastructure.

To transform the property market into an integrated, efficient and user-friendly digital ecosystem, a comprehensive approach is necessary. This approach must address all aspects of the exchange, security, standardization and scalability of data. It must go beyond the technical implementation of APIs and consider the broader context of market dynamics, regulatory requirements and user needs.

A well-designed PMI should provide a framework for secure and seamless data exchange, ensuring that all participants in the property market can interact and transact with confidence. It should establish common data standards and protocols that enable interoperability and data reusability across different platforms and systems. Moreover, it should incorporate intelligent mechanisms for data synchronization, conflict resolution and real-time updates, ensuring that all parties involved in a transaction have access to accurate and up-to-date information.

A blueprint for a property market infrastructure

The success of global financial markets can be largely attributed to the development of robust and efficient FMIs such as Swift, Visa and Mastercard. These infrastructures have revolutionized the way financial transactions are conducted, making them more secure, transparent and accessible. By drawing on the key principles and strategies employed by these FMIs, we can create a blueprint for a PMI that has the potential to transform the way property transactions are carried out.

At the heart of the proposed PMI is a standardized identity framework, which ensures that all actors on the network are known and trusted entities. This framework is essential for building trust and confidence in the property market, as it allows for secure and transparent interactions between buyers, sellers, estate agents, banks and legal entities. By implementing a robust identity-verification process, the PMI can mitigate the risks of fraud and money laundering, which are significant concerns in the property market.

Figure 1. The proposed property market infrastructure.

The PMI would be able to provide a common API that allows participants, when they are connected to the network, to connect to all other participants. This 'connect to one, connect to many' capability streamlines the integration process, allowing stakeholders to easily access and share information across the network. But there is much more to a PMI.

Another key aspect would be ensuring that the data passing through the PMI was auditable and immutable. This would provide a clear and tamper-proof record of all property-related activities and ensure trust in data. In addition, the PMI would also offer a user-friendly interface for non-API access. This user interface would allow stakeholders who may not have the technical expertise or resources to integrate with the API to still participate in the network. The interface would provide a secure and intuitive platform for managing property transactions, from listing properties to finalizing deals. This inclusive approach ensures that all stakeholders, regardless of their technical capabilities, can benefit from the efficiency and transparency offered by the PMI.

The standardization of processes and data formats is another crucial aspect of the PMI. By establishing a universal set of standards for property transactions, the PMI would enable seamless communication and data exchange between all participants in the network. This standardization not only reduces the time and costs associated with property transactions but also enhances the overall efficiency of the property market.

Moreover, by providing a clear and consistent framework for property transactions, the PMI would help to reduce the potential for disputes and legal challenges, further strengthening trust in the market.

The network effect would be a powerful driver of adoption and growth for the PMI. As more stakeholders join the network, the value and utility of the PMI increases exponentially. This network effect creates a virtuous cycle, in which the increased participation of buyers, sellers, estate agents, banks and legal entities leads to improved market transparency and enhanced opportunities for collaboration and innovation. With the growth of the network, the PMI would become a valuable tool for anyone looking to participate in the property market.

Openness and interoperability are fundamental to realizing the full potential of the PMI. By ensuring that PMI standards are universally accepted and compatible with existing systems and processes, the infrastructure could facilitate seamless transactions and enable the participation of a wide range of stakeholders. This universal compatibility also allows for the integration of the PMI with other critical infrastructures and existing systems, such as the Land Registry, financial systems and estate agents' customer-relationship management (CRM) and case management systems.

The ownership model of the PMI would follow the successful examples set by Visa, Mastercard and Swift. These FMIs were initially invested in by large market incumbents that recognized the need to collaborate on a shared market infrastructure. By involving key stakeholders in the property market as investors and owners of the PMI, the infrastructure can benefit from their expertise and resources. This collaborative approach ensures that the PMI is developed and governed in a way that addresses the needs and concerns of all participants in the property market, fostering widespread adoption and trust in the infrastructure.

To stay at the forefront of innovation and respond to the evolving needs of the property market, the PMI would need a culture of continuous improvement. This could involve the incorporation of emerging technologies such as blockchain, smart contracts and AI to further enhance the security, efficiency and transparency of property transactions.

Case study: Coadjute
In 2018 I had the privilege of leading the innovative Land Registry Digital Street blockchain project. This was an R&D project to explore whether blockchain could provide a market infrastructure to facilitate end-to-end property transactions.[7] The project was a success and reinforced the need for a PMI. However, there was no government money available to continue the project and build a live version of the network. Recognizing the critical need for such an infrastructure and its immense

potential, my co-founders and I raised venture capital and founded Coadjute.

Coadjute is an open network designed to connect digital platforms and services, enabling estate agents, buyers, sellers, banks, building societies, lawyers and other stakeholders to collaborate seamlessly during property transactions. The Coadjute network is a prime example of a PMI that follows the blueprint of successful FMIs such as Visa, Mastercard and Swift by having strategic investment and board membership from the largest market participants: Rightmove, Lloyds Banking Group, Nationwide and NatWest.

At the core of the Coadjute network is a transaction orchestration service. Orchestration is the process of efficiently coordinating the various tasks, activities and events in a property transaction in order to bring together the different business processes of the involved parties into a seamless digital journey.[8] Coadjute's goal in transaction orchestration is to ensure smooth and synchronized operations throughout the transaction, eliminating the complexities and inefficiencies of manual processes.

By integrating systems and data, Coadjute enables real-time information exchange, automated workflows and better visibility of the transaction progress. This leads to improved efficiency, fewer errors and increased transparency. Importantly, the process of orchestration does not attempt to replace existing systems or processes or try to centralize the whole industry with a monolithic database. Instead, through digital connectivity, Coadjute enables seamless coordination and collaboration, ensuring consistent end-to-end communication.

Coadjute enables interoperability through a standard, industry-wide data model and sophisticated APIs that connect to legacy systems. The network translates between these legacy systems, enabling them to interoperate and communicate securely and seamlessly and introducing the completely novel capability of true multi-party collaboration.

In essence, Coadjute simplifies the property transaction process by harmonizing the tasks of all parties involved, resulting in a more efficient and transparent experience for everyone.[9] As the property industry continues to evolve, other initiatives like Coadjute will play a crucial role in shaping the future of property transactions.

Conclusion

The potential transformation of the UK property market can take inspiration from global finance and the shared infrastructures exemplified by Visa, Mastercard and Swift. These models demonstrate how cooperative frameworks and technology can revolutionize complex industries.

In the UK, with its annual property asset exchanges exceeding £300 billion, the absence of a unified market infrastructure is not just an operational challenge but a major economic bottleneck. The current electronic market is characterized by fragmented, siloed systems, in stark contrast to the efficiency seen in cash transactions. Houses, being unique and non-fungible, necessitate a robust, streamlined mechanism for transaction management, a role that a well-designed PMI could fulfil.

The Land Registration Act 2002 lays the groundwork for such an infrastructure, envisioning a digital Land Registry Network akin to FMIs in finance. However, this vision remains underrealized, which points to a disconnect between legislative ambitions and their practical execution.

Coadjute's emergence as an open market infrastructure signals the scope for a forward leap in property transactions. Inspired by the Land Registry's blockchain pilot, Coadjute's goal is to orchestrate seamless interactions between all parties in property transactions. This integration promises to dismantle the complexities of manual process, boosting efficiency, transparency and the overall quality of the transaction experience.

By providing a standardized identity framework, a common API, data integrity and auditability, and a user-friendly interface, a PMI can create a trusted, efficient and inclusive ecosystem for property transactions. Network effects, interoperability, a collaborative ownership model and a culture of innovation will be critical drivers of the PMI's success and can ensure that it remains a vital tool for unlocking the full potential of the property market in the years to come. This PMI can transform the property market into a dynamic, efficient and user-friendly landscape, supporting the UK's economic growth and serving consumer and business needs.

CHAPTER 6

Data standards

> The true value of data can only be fully realised when it is fit for purpose, recorded in standardised formats on modern, future-proof systems and held in a condition that means it is findable, accessible, interoperable and reusable. By improving the quality of the data, we can use it more effectively, and drive better insights and outcomes from its use.
>
> — National Data Strategy, 2020[1]

In this chapter I look at the importance of standardized data for an effectively functioning property market. Data acts as the market's lifeblood, crucial for its efficiency. However, if I were to assess the property market's data maturity using the government's data maturity framework, which grades from beginner (1) to mastery (5), I would classify it firmly at the beginner level.[2] This status is marked by a reliance on external mandates for implementations of data standards and a focus on interoperability solely to prevent critical system failures.

This analysis sets the stage for a deeper exploration into the current use of data in the property market, pinpointing the primary obstacles that hinder progress. By drawing parallels with the financial services sector, where data fluidity between institutions is the norm, I aim to uncover actionable insights.

For those interested in what is happening in the data space, I also cover key government strategies, legislative efforts and industry initiatives that are pushing towards a fully digital property market based on open data.

Data is essentially information transformed into a format that is easy to move and process. It is made up of facts – these could be numbers, words, measurements, observations or simple descriptions. When we talk about a property, the specific details about it – such as the title number, address, owner and energy rating – are its 'attributes'. These attributes are different pieces of data that give us valuable insights into the property.

The problem with data today

In the property market, the way we handle and understand data is fundamental to the efficiency and success of transactions. A property transaction involves multiple participants, including buyers, sellers, estate agents, conveyancers, mortgage brokers and lenders, each facing unique challenges stemming from the current state of data management. To assess the quality and effectiveness of data management in the property market, we can compare it against the FAIR data principles (Findable, Accessible, Interoperable and Reusable), which are the ideal standard for any market.[3] Examining the property market's data through this lens, we soon see that it falls short of being truly FAIR.

One of the primary issues with data in the property market is that it is often hard to find. Both consumers and professionals struggle to locate the necessary data at the right time in order to make informed decisions. This lack of findability can lead to delays in the transaction process, as parties spend valuable time and resources searching for the information they need.

Even when data is located, accessing it can be slow, difficult and costly. Whether the problem is obtaining information from local authorities and management agents, or simply gathering upfront property details from a seller, the process is fraught with hurdles. This lack of accessibility not only frustrates consumers

DATA STANDARDS 71

but also hinders the ability of professionals to provide efficient and effective services.

Another major challenge in the property market is the lack of interoperability between systems. Despite the presence of numerous APIs, easy data transfer between multiple services is still rare. Systems are not designed to be interoperable, meaning that cumbersome, bilateral efforts are needed to exchange information. This lack of seamless data exchange can lead to errors, duplicated work and a general slowdown of the entire transaction process.

On top of these obstacles to accessing and exchanging data, re-using it across the property market is also difficult due to the absence of a universal data standard. While data can be transformed across systems, the lack of adopted standards means that data transformations must be conducted on a partner-by-partner basis. This process complicates workflows and increases the risk of data quality issues and transmission errors.

The consequences of these shortcomings in data management extend beyond inconvenience. For consumers, the difficulty in finding, accessing and understanding the relevant data can lead to ill-informed decisions, increased stress and a general sense of frustration with the process of buying or selling property. This negative experience can erode trust in a transaction and ultimately result in transactions falling through.

For businesses operating in the property market, the challenges posed by poor data management can be equally severe. Estate agents, conveyancers, mortgage brokers and lenders all rely on accurate, timely and comprehensive data to provide their services effectively. When data is hard to find, access or re-use, these professionals are forced to spend more time and resources on administrative tasks related to manual data handling and verification, rather than focusing on delivering value to their clients. This not only reduces their productivity but also increases the cost of their services, which is ultimately passed on to consumers.

Moreover, the lack of interoperability and standardization in data management can hinder innovation and competition in the property market. When data is siloed and difficult to exchange, new entrants can find it challenging to develop and offer innovative services that could benefit consumers and improve the overall efficiency of the market. Indeed, there are large players in the market whose business model is based on preserving data silos, which has led to stagnation in parts of the market, with established players facing little pressure to improve their practices or innovate.

Case study: the role of data standards in financial markets – Swift and ISO 20022

This case study explores the transformative impact of Swift and ISO 20022, two pivotal standards that have reshaped the way financial information is exchanged globally.

The origins of Swift (outlined in chapter 5) can be traced back to the 1970s, a time when the financial world was in dire need of a system that could simplify and secure the way financial institutions communicated globally. Swift emerged as that much-needed solution, offering a standardized messaging network that linked banks and financial institutions worldwide. This was a significant leap forward from the disjointed and often unreliable communication methods previously in place.

In tandem with the evolution of Swift, ISO 20022 was developed as a universal standard for the electronic exchange of data. Its goal was straightforward: to bring uniformity to the various data formats that had been used in financial transactions. The shift to ISO 20022 represented more than just an upgrade; it was a move towards a more cohesive system for financial communication.[4]

The implementation of Swift and ISO 20022 fundamentally changed the financial sector. For starters, the standardization of messaging formats introduced a level of clarity and consistency that was previously unheard of. The confusion and misunderstandings that once complicated financial transactions were largely eliminated.

These standards significantly boosted efficiency. With Swift and ISO 20022, transaction times were shortened and the errors associated with manual processing were almost eradicated. This not only benefitted financial institutions but also markedly improved the experience for customers engaging with financial services.

Perhaps one of the most transformative effects was on international transactions. Swift and ISO 20022 effectively bridged the divide between different financial systems, making it easier and faster to conduct transactions across borders. They also enhanced security measures that safeguarded the privacy and integrity of financial data.

The integration of Swift and ISO 20022 into financial operations led to cost reductions and process optimization, making financial services more reliable and accessible. Moreover, compliance and regulatory reporting, which used to be labour-intensive, became more manageable thanks to the consistency in data communication.

Property market data standards

The quest to establish data standards in the UK property market has recently seen significant progress. I have been closely involved in these efforts, and it has been incredibly rewarding to witness the results when the industry and government collaborate effectively. The bottom line is that the UK is on a promising trajectory towards realizing property data standards and reaping their benefits. This movement is primarily driven by three pivotal initiatives.

Firstly, strong government leadership has been instrumental in paving the way forward. Initiatives such as the National

Data Strategy and supporting legislation have provided a clear direction and framework for the development and adoption of data standards in the property market. This high-level support has been crucial in creating an environment in which industry stakeholders feel empowered and motivated to work together towards a common goal.

Secondly, the power of industry collaboration cannot be overstated. The collective efforts of the Home Buying and Selling Group, the Open Property Data Association, the Law Society and the Digital Property Market Steering Group have fostered a spirit of cooperation and shared purpose, ensuring that the development of data standards is a truly collaborative process.

Thirdly, the increasing adoption of emerging industry data standards by businesses is a clear sign that the tide is turning. As more and more companies recognize the benefits of standardized data – such as improved efficiency, reduced costs and enhanced customer experiences – they are beginning to incorporate these standards into their own processes and systems. This growing momentum is creating a positive feedback loop, since the more businesses adopt the standards, the more valuable and compelling it becomes for others to follow suit.

One of the key advantages of widespread business adoption is that it helps to create a critical mass of standardized data within the property market. This, in turn, makes it easier for other stakeholders – such as government agencies, regulators and consumers – to access and utilize this data for their own purposes. For example, standardized property data could be used to inform policy decisions, support market analysis or empower consumers to make more informed choices when buying or selling a home.

There are still technical, legal and cultural hurdles to overcome, and it will take sustained effort and commitment from all stakeholders to see this vision through to fruition. However, the progress made to date is highly encouraging, and there is a growing sense of optimism within the industry and a clear determination to keep pushing forward.

Strong government leadership

The UK government's dedication to fostering a digital economy, with data as a pivotal component, is reshaping numerous sectors, including the property market. Three key initiatives – namely, the National Data Strategy, the Data Protection and Digital Information Bill and the Smart Data Working Group – are instrumental to this effort, which has recently received further support from the government's smart data roadmap.

National Data Strategy

In 2020 the UK government unveiled the National Data Strategy, a pivotal blueprint for transforming various sectors, including the property market.[5] The strategy recognizes data as the core of modern economies, essential for fostering innovation and improving operational efficiencies.

At the heart of this strategy lies the concept of data foundations, which is particularly relevant to the property market. This concept underscores the need for data to be appropriate, standardized and stored in systems that are adaptable to future advances. For the property market, this signifies a shift from disjointed, archaic systems to integrated, digital platforms where data is readily accessible, interoperable and reusable.

The National Data Strategy advocates for the optimized use of data to enhance service delivery, spawn new products and propel technological innovation. In the property market this translates to data-driven transactional processes, improved client services and the birth of novel business models. The strategy envisions a property market bolstered by a robust data infrastructure and a fresh perspective on data management and sharing.

The strategy proposes the comprehensive integration of data usage across various sectors, including business, government and civil society. For the property market, this translates to improved cooperation and data sharing among different stakeholders, such as agents, brokers, conveyancers, mortgage lenders, HM Land Registry and local authorities. This integrated approach is geared towards creating a more cohesive and user-friendly market experience.

The National Data Strategy lays down a solid framework and a supportive policy environment for transitioning the property market from a market with an electronic base to one that is truly digital. It underscores the significance of using data as a catalyst for innovation, operational efficiency and economic expansion.

Data Protection and Digital Information Bill

At the time of writing, the property market is on the brink of a pivotal transformation thanks to the Data Protection and Digital Information Bill, which is due to get Royal assent in 2024.[6] This legislation is poised to usher in a new era of compulsory data sharing within the property sector.

This journey towards smart data legislation began back in 2018, when the government set out to extend the benefits of smart data beyond its initial success in the financial sector with the Open Banking initiative. The goal of this initiative was to make financial data more accessible and manageable, a model the government hoped to replicate across other sectors. The feedback from these explorations was clear: smart data held immense promise, but achieving its full potential across various industries would require legislative backing.

The resulting Data Protection and Digital Information Bill is much more than a mere legal requirement; it represents a strategy to tackle the persistent data challenges that hinder the growth of the digital economy. The bill grants the government the authority to establish and enforce data standards, promoting uniformity and consistency in data management across different sectors. In particular, the authority to mandate that data holders conform their APIs to certain standards is expected to be key in addressing data challenges and diminishing information asymmetry in the digital economy.

By legislating the standardization of data sharing and management, the bill lays the foundation for a property market that operates more efficiently and transparently. It promises to streamline the flow of information, allowing for more informed decision making and smoother property transactions.

Once enacted, the Data Protection and Digital Information Bill will mark a major milestone towards achieving a data-driven and integrated industry. This move towards a smart data framework is set to redefine the digital economy and the property market, and it underscores the transformative power of legislative support in realizing the full potential of smart data.

Smart Data Working Group

The Smart Data Working Group marks a significant move by the government towards building a data-empowered economy.[7] It brings together experts and representatives from key organizations such as HM Treasury and the Department for Digital, Culture, Media and Sport, alongside others such as Ofcom and the Financial Conduct Authority.

The vision at its core is to spread the advantages of smart data across different markets, sparking an era of new services, competitive pricing and broader choices for consumers and small businesses.

In terms of its impact on the property market, the potential for a smart data scheme is immense. It promises to smooth out the data exchange between estate agents, conveyancers, mortgage brokers and lenders, making transactions more efficient, reducing red tape and improving the overall buying and selling experience.

The government's smart data roadmap up to 2025

The chancellor's 2023 autumn statement outlined the government's ambition to initiate a smart data revolution, which could potentially increase the UK's gross domestic product (GDP) by £27.8 billion annually. Data portability through smart data will play a crucial role in driving this growth. The data economy is a significant and expanding part of the overall economy, with the Organisation for Economic Co-operation and Development estimating that it accounts for 3–6.7% of UK economic activity.

The government's smart data roadmap outlines plans to explore the application of smart data powers – as defined in the Data Protection and Digital Information Bill – across home buying and six other sectors: energy, banking, finance, retail, transport and telecommunications.[8] The roadmap aims to expand on the principles that have made Open Banking an international success and apply them to the broader UK economy. For each sector, the government intends to identify areas where regulatory intervention may be necessary, consult on the use of the smart data powers outlined in the bill and, if necessary, design a potential scheme before proceeding with implementation. By following this approach, the government seeks to unlock the potential of smart data and drive economic growth across various sectors of the UK economy.

Industry collaboration on data standards

With the government leading the charge towards an open data economy, the industry is not far behind in developing data standards. These standards are crucial as they significantly enhance the value of data. As highlighted by PwC's 'Putting a value on data' report, industry collaboration on data standards not only facilitates interoperability and efficiency but also directly contributes to the economic value of data by making it more accessible and usable across the property transaction.[9]

There is wide consensus across government and industry that data standards would be a good thing for the property market. Two key industry bodies at the forefront of this development are the Home Buying and Selling Group and the Open Property Data Association, while the Law Society, through its digitalization of transaction forms, is also contributing to the conditions for greater alignment on data standards.

Home Buying and Selling Group

The Home Buying and Selling Group is a coalition of public and private sector organizations from various corners of the property industry, united in the goal of enhancing the home buying and selling experience for property owners.[10]

I became a member of this group about four or five years ago, as I saw its potential to drive significant change in the property market. A notable achievement of our group has been the creation of the Buying and Selling Property Information (BASPI) form.[11] This initiative represents a significant move towards a universal data standard for the property market, and it is an essential step in our ongoing quest to streamline and improve the property transaction process.

Our approach was thorough and involved dissecting the BASPI form alongside other critical industry documents to pinpoint unique data points. This endeavour led to the unveiling of the first version of the data standard schema and API. Made available on GitHub under an open-source licence, these tools represent a significant step towards establishing an open, accessible and standardized data framework for the property market.[12]

Since its introduction, the schema has seen over 180 updates to the GitHub repository, reflecting the standard's dynamic and evolving nature. Its growing adoption among businesses signals a shift towards industry-wide normalization.

The rollout of this data standard is a landmark achievement for the property industry. It is designed to streamline data management, boosting efficiency and reliability. But it is more than an improvement on existing processes; it sets the groundwork for future innovations and ensures that the property market remains flexible and in tune with changing demands.

Open Property Data Association

The recently formed Open Property Data Association (OPDA) is setting the stage for a transformation in how property data is shared and used across the industry.[13] The OPDA brings together technology experts and business leaders from various sectors of the property market, including portals, lenders, conveyancers and brokers.

The group's founding principle is to make property-related data standards open-source and freely accessible to all. It is committed to maintaining the open standards that its members develop as a free resource for the industry, supporting their continued development and ensuring they remain managed as an open-source project. The association seeks to lead the way in promoting the adoption of trusted open property data by offering a platform for collaborative innovation that promises to benefit the entire property industry.

Law Society transaction forms

Law Society transaction forms are fundamental to how conveyancing lawyers handle property transactions.[14] These forms dictate the necessary information at different stages of a transaction and for various transaction types. For example, the TA6 form, completed by the seller early in the transaction, communicates crucial property information to the buyer.

These forms are now moving from physical documents to a data-driven format, and the transition promises to streamline the conveyancing process significantly. Furthermore, aligning these digital forms with open standards, such as those promoted by the OPDA, ensures their interoperability within the broader property market ecosystem.

Why collaborate?

In the drive to transform the property market, the push for open industry standards has emerged as a cornerstone of innovation. The rationale behind this approach is simple yet profound: proprietary standards create isolated islands of information, hindering seamless communication across the sector. It is akin to crafting a unique language that is decipherable to only a select few and therefore limited in its utility and adoption.

My personal commitment stems from a deep-seated belief in the transformative power of open data standards. The driver for participation in these open projects is an understanding that true value lies in universal accessibility and comprehension.

This commitment is not solely for the benefit of service providers or industry insiders; it is ultimately about enhancing the consumer experience. For prospective homeowners, the move towards open standards means a more transparent, accessible and navigable path to securing their dream home. Open standards promise to demystify the complexities of property transactions, making the market more approachable for everyone involved.

Conclusion

The current state of data in the online property market lacks standards, and its fragmented and siloed nature creates a complex, disjointed process for buyers and sellers. If we want a digital property market, the importance of FAIR data principles is therefore undeniable. Their adoption promises a future in which transactions are streamlined and all market participants, especially consumers, can benefit from enhanced transparency and efficiency.

Government initiatives such as the National Data Strategy, coupled with industry efforts such as the Home Buying and Selling Group and the OPDA, signify a collective movement towards a digital property market underpinned by robust data standards. These collaborations, emphasizing the need for open and standardized data, pave the way for a more interconnected and efficient market that can address long-standing challenges and unlock new opportunities for innovation and growth.

As we all push towards the digital property market, my call to action for industry leaders and government is to embrace the digital transformation with a unified commitment to data excellence.

CHAPTER 7

Digital trust: the golden thread

> Following the Grenfell tower tragedy, the government appointed Dame Judith Hackitt to lead an independent review of building regulations and fire safety. In her report, 'Building a safer future', Dame Judith recommended the introduction of a 'golden thread' as a tool to manage buildings as holistic systems and allow people to use information to safely and effectively design, construct and operate their buildings. She stated that 'a robust golden thread of key information' should be 'passed across to future building owners to underpin more effective safety management throughout the building life cycle'. Dame Judith set out that these recommendations were to address the problems within the industry of key information not being effectively managed, or even available throughout the life cycle of the building, including when there is a change in ownership.
>
> — Building Regulations Advisory Committee, 2021[1]

I remember reading the final report of the Independent Review of Building Regulations and Fire Safety in May 2018 and being struck by the concept of a 'golden thread' of data throughout a building's life cycle.[2] I thought that what the property market needed was just such a golden thread running through the property transaction process. It seemed to me that the creation of a verified and trusted flow of data from the start of the property transaction at listing right through to completion and the handover to the new owner would completely transform the transaction process.

When I think about the potential of a golden thread of data for each property transaction, I envision a future in which transactions are seamless, transparent and built on a foundation of trust. A future in which buyers, sellers and professionals can access reliable, up-to-date information at their fingertips, without the need for endless verification and duplication of efforts.

In this chapter I explore the current challenges around data trust that are plaguing the property market, and I investigate how the golden thread concept can address these issues. Drawing insights from the Building Regulations Advisory Committee's 'golden thread' report, I look at the principles that underpin the creation of a data golden thread.

However, the property transaction golden thread is more than just a theoretical concept. This chapter also examines the practical steps being taken to make it a reality, from the introduction of the digital identity and attributes trust framework, to the emergence of 'digital data packs' and the pivotal role of the Law Society transaction forms.

As you navigate this chapter, you will discover how the PTGT can benefit all stakeholders in the property market, from buyers and sellers to conveyancers, lenders and estate agents. You will see how it can reduce duplication of effort, improve data quality, enhance transparency, increase efficiency and ultimately lead to a better user experience for everyone involved.

The problem with data in today's transactions

In today's property market, professionals are inundated with a constant stream of information. From an ever-increasing list of material property information, to client data and transaction records, they must navigate a complex web of data spread across multiple locations and formats. A significant part of their job involves continuously checking, double-checking and verifying this information to ensure accuracy and consistency.

Imagine a typical property transaction involving multiple parties, such as estate agents, conveyancers, brokers, lenders, surveyors and HM Land Registry. Each of these parties plays a crucial role in the process, but a fundamental issue arises: they do not fully trust each other – or at least not the data flowing between them. This lack of trust leads to a time-consuming and costly process of repeatedly verifying information, whether that is property information, transaction information or identity information, to name but three.

Consider a property's title number, a unique identifier that links a property to its record in the Land Registry. In the current system, each party in the transaction independently verifies this number. The estate agent checks it, the conveyancer confirms it, the lender verifies it, and so on. This duplication of effort stems directly from the lack of a trusted data exchange mechanism.

The consequences of this mistrust are far-reaching, impacting the operational efficiencies of all organizations involved in the transaction. The Land Registry, for example, reports that nearly 30% of submissions – which come at the final stage of a transaction – fail their data-checking process. This means that even after all the double-checking and verification before the data reaches the Land Registry, almost a third of the applications still contain errors or inconsistencies. This not only wastes time and resources but also frustrates all parties involved.

The cause of all this rechecking

But why do property professionals feel compelled to check information provided to them by other professionals? In the physical world, trust is built through face-to-face interactions and tangible evidence. For example, when you meet a client in person, you can verify their identity by examining their ID or recognizing them from previous meetings.

In the online world establishing trust is more complex, because the data (the information or content) is often separated from its source (the person or entity that created or provided the data). This separation makes it difficult to verify the origin and authenticity of the information, as there is no direct, observable link between the data and its creator. This disconnect between data and source creates several challenges in establishing trust and reliability in a digital context.

Firstly, identity confirmation in the digital world is not straightforward. Without the ability to physically examine identification documents or rely on personal recognition, verifying the identity of the person providing the data becomes a complex task.

Secondly, even if you can confirm the identity of the person providing the data, it is equally important to establish a strong link between the data and the verified identity. In other words, you need to be able to prove that the specific piece of information came from the identified individual. This binding of data to its source is crucial for establishing trust in the information.

Thirdly, as data passes between organizations and systems, maintaining the connection between the data and its original source becomes increasingly difficult. Each time the data is transferred or processed, the risk of the data–identity link being broken or compromised increases.[3]

The result of these challenges is that property professionals often resort to manually confirming all data themselves, going back to the original sources to verify the information. Imagine an estate agent sending a sales memorandum to a conveyancer that includes confirmation of title and the property's ownership. In the online market the conveyancer may have doubts about the authenticity of the email and the accuracy of the information provided. They may question where the information came from and whether it had been correctly verified. To alleviate these concerns, the conveyancer may feel compelled to log into the Land Registry to directly verify the information, despite the fact that the estate agent had got all the information from the Land Registry in the first place.

The lack of inherent trust in data within the property ecosystem and the difficulty in maintaining the connection between data and its source throughout the transaction process are at the root of the inefficiencies and delays plaguing the digital property market. Until we can find ways to establish and preserve the data–identity link, property professionals will continue to feel the need to manually verify information, hindering the speed and efficiency of transactions.

This problem is not unique to the property market; it is a fundamental challenge faced by many industries in the digital age.[4] As more and more of our interactions and transactions move online, the need for robust digital trust frameworks becomes increasingly pressing. Initiatives such as digital identity schemes, blockchain technology and secure data sharing protocols are all aimed at addressing this issue, seeking to create a digital environment in which data can be trusted and relied on without the need for constant manual verification.[5]

A golden thread of data

In the wake of the tragic Grenfell Tower incident, Dame Judith Hackitt's Independent Review of Building Regulations and Fire Safety introduced a groundbreaking concept: a 'golden thread' of information. The concept highlights the importance of maintaining a continuous thread of data throughout a building's life cycle.

The golden thread provides a single, reliable source of information that captures all relevant data about a building, from its design and construction to its ongoing maintenance and operation. This information is digitally recorded, easily updatable and accessible to all relevant stakeholders, ensuring that everyone involved in the building's life cycle has access to accurate and up-to-date information.

By maintaining this golden thread of data, building owners, managers and occupants can make informed decisions, so that the building remains safe, compliant and fit for purpose throughout its life cycle. This approach promotes transparency, accountability and collaboration among all parties involved in the building's management.

The property transaction golden thread

Just as the review's proposal of a golden thread of data ensures the safety and efficiency of a building throughout its life cycle, a PTGT can streamline the process of buying and selling property, making it more transparent, secure and efficient.

The Building Regulation Advisory Committee has set out ten principles for the golden thread.[6] These outline the importance of accurate and trusted information, residents feeling secure in their homes, culture change, a single source of truth, secure data, accountability, understandable and consistent information, simple access, the longevity and shareability of information, and relevant and proportionate data. The ten principles are detailed below in the context of the PTGT.

1. *Accurate and trusted.* The golden thread must be accurate and trusted so that the relevant people use it. The information produced will therefore have to be accurate, structured and verified, requiring a clear change-control process that sets out how and when information should be updated and who is responsible for updating and checking it.

2. *Clients secure in transactions.* Clients should be provided with accurate information in order to hold property professionals accountable and ensure efficient, compliant transactions.

3. *Culture change.* The golden thread should support culture change within the industry, as it will require increased competence and capability, different working practices, updated processes and a focus on information management and control. The golden thread should be considered an enabler of better and more collaborative working.

4. *A single source of truth.* The golden thread should bring all information together in a single place, meaning there is always a 'single source of truth'. It will record changes

(i.e. updates, additions or deletions to information, data, documents and plans), including the reason for the change, the evaluation of the change, the date of the change and the decision-making process. This will reduce the duplication of information (doing away with email updates and multiple documents) and help drive improved accountability, responsibility and a new working culture.

5. *Secure.* The golden thread should protect and control access to personal information in order to maintain transaction security and comply with the General Data Protection Regulation (GDPR).

6. *Accountable.* Changes should be recorded and clear duties should be set for all parties in order to maintain information standards and drive accountability.

7. *Understandable and consistent.* Information should be clear, focus on user needs and use standard methods and terminology to allow for a consistent understanding.

8. *Simple to access.* The golden thread should store information in a structured, easily accessible way, guided by digital standards.

9. *Longevity, durability and shareability.* Information should be formatted for long-term maintenance, interoperability and sharing among parties using different software.

10. *Relevant and proportionate.* Information should be regularly reviewed and updated to ensure relevance to the transaction, with unnecessary data being discarded.

Building a golden thread of property transaction data may seem like a distant goal, but recent developments suggest that it is closer to reality than many might think.

The UK is introducing a new system called the digital identity and attributes trust framework,[7] which is based on a new law called the Data Protection and Digital Information Bill (discussed in more detail in chapter 6).[8] The bill is expected to receive official approval in 2024, and it is designed to create a trusted environment in which businesses can securely share data within their industries.

To test the feasibility of such an environment, the government has already conducted trials using a preliminary version of the digital identity and attributes trust framework. The framework enables the secure and verified sharing of digital identity data between three key parties: identity service providers (those who create the data), data subjects (those whom the data is about) and relying parties (those who wish to use the data). The successful implementation of this framework in the property market could significantly streamline the process of buying and selling property.

As the UK moves closer to officially approving the Data Protection and Digital Information Bill, the reality of a golden thread of property transaction data becomes increasingly tangible.

ToIP

The Trust over IP (ToIP) framework utilizes a technology stack that enhances security, transparency and efficiency when applied to property transactions.[9] At the foundation of this stack is the public utilities layer, which consists of distributed ledger technology (DLT) and a decentralized public key infrastructure (DPKI). In a property transaction, a DLT, such as a blockchain, creates an immutable record of the transaction and the associated digital identities, ensuring transparency and preventing tampering. The DPKI is used to create and manage the cryptographic keys for the digital identities of all parties involved in the transaction, providing a secure foundation for trust.

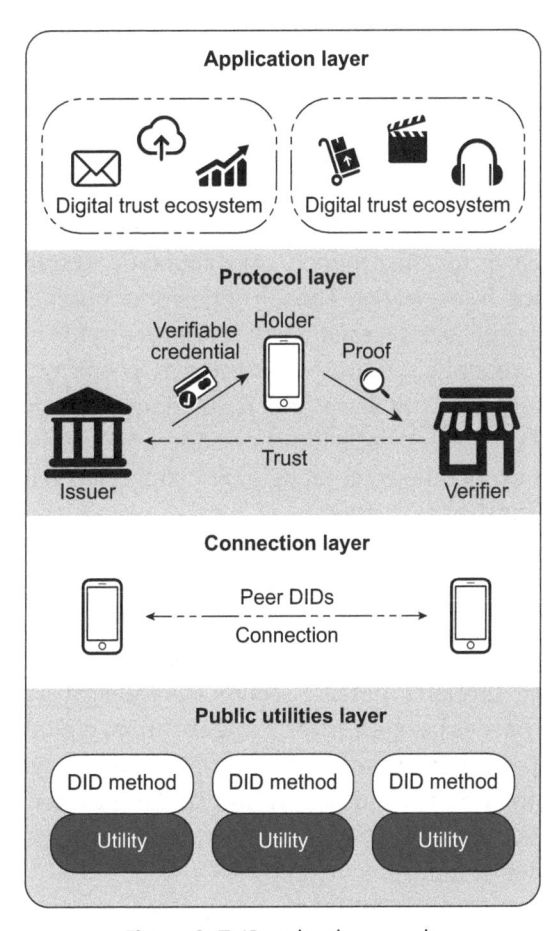

Figure 2. ToIP technology stack.

Building on this foundation is the connection layer, which includes the exchange of decentralized identifiers (DIDs) and verifiable credentials (VCs). Each party in the property transaction, including the buyer, seller, estate agents and banks, would have their own DID, which is a unique, cryptographically-verifiable identifier. DIDs allow for secure, private and decentralized control over digital identities. VCs are digital attestations of qualifications, competencies or other attributes. In a property transaction, VCs could include proof of property ownership for the seller, proof of identity and creditworthiness for the buyer, and professional licences for estate agents. VCs can be securely issued, held and verified using the ToIP protocol layer.

The protocol layer of the ToIP stack defines the rules, policies and standards for the property transaction ecosystem through governance frameworks. These frameworks outline the roles and responsibilities of each party, the process for issuing and verifying credentials, dispute-resolution mechanisms, and compliance with relevant laws and regulations. The governance frameworks ensure that all parties adhere to the same set of rules and expectations, creating a consistent and reliable environment for the transaction.

Finally, the application layer consists of the specific applications and services built on top of the ToIP stack for property transactions. These ecosystem-specific applications could include property listing platforms that use VCs to verify ownership and property details, secure communication channels between parties using DIDs for authentication, digital-contract-signing applications that leverage DIDs for identity verification and VCs for proof of authority, and secure payment systems that use VCs to verify the buyer's financial status and facilitate the transfer of funds.

By leveraging the ToIP technology stack, a property transaction can be made more secure, transparent and efficient. The combination of DIDs, VCs and governance frameworks creates a trusted environment in which all parties can verify the identities and qualifications of the others, reducing the risk of fraud. The application layer builds on this trust to provide ecosystem-specific solutions that streamline the transaction process and improve the user experience. The ToIP framework has the potential to transform the way property transactions are conducted, making them faster, safer and more accessible to all parties.

Building trust with upfront information: the first step in creating the property market's golden thread

Providing 'upfront information' – which is the concept of ensuring that a seller discloses all material information about a property to a potential buyer at the point of listing the property for sale – has long been a major topic in the property market. Despite efforts from the industry and government, a solution has not yet been found, mainly because the property market lacks easily accessible, trusted data.

The problem is that, even if data is collected at the listing stage, it is not stored in modern data systems and does not have the necessary details about its source. This means it cannot act as a golden thread of data that follows the transaction from start to finish. Below, we will look at the background of this challenge.

Home information packs

First proposed in the UK by the Labour Party in the run-up to the 1997 election, home information packs (HIPs) were intended to reform the home buying process and tackle gazumping (when a seller accepts a higher offer after already accepting one from another buyer). After winning the election, the Labour Party found that gazumping was not as common as it thought, affecting fewer than 1% of transactions. It then gathered feedback from over 2,000 consumers and industry experts to pinpoint the major issues affecting the buying and selling experience.

A big problem it found was that important information was given to buyers too late. The proposed HIPs were now presented as a way to fix this by providing key documents – such as property searches, title evidence and energy performance certificates – early on. The goal was to speed up transactions, with the additional benefit of reducing the chance of gazumping and gazundering (when a buyer lowers their offer just before the exchange of contracts).

HIPs faced a lot of criticism and were eventually suspended by the coalition government in 2010, with only the need for energy performance certificates remaining. This left the original problem unsolved.

Digital data packs

Fast forward to the 2020s, and the government seems committed to a digital version of HIPs, as stated in its Levelling Up white paper.[10] It wants to work with the industry to make sure buyers can digitally access critical information from trusted sources.

In November 2023 the National Trading Standards Estate and Letting Agency Team released new guidelines to help protect consumers and ensure essential property information is available upfront. It divided this information into three parts.

- Part A: always necessary information, such as council-tax band, asking price and property tenure.

- Part B: information about the property's physical characteristics, maintenance costs, mortgage and insurance availability, and aspects affecting its use or enjoyment.

- Part C: information that may be necessary depending on the property's specific circumstances and location (the guidelines advised estate agents to seek professional advice for matters such as building safety, property restrictions, flood risks and planning permissions).[11]

These guidelines aimed to increase transparency in the property market and help consumers make informed decisions when searching for a property.

The industry is now responding to this heightened attention from both the government and Trading Standards regulations. We are witnessing the emergence of upfront 'digital data packs'. Unlike their predecessors, which consisted of static information, these data packs are built on open data standards and ensure that every piece of data comes from a verified source and is cryptographically bound to its provenance. This not only enhances the utility, reusability and interoperability of property information but also ensures that data gathered at the start of the transaction can be trusted and utilized by all parties throughout the process, creating a PTGT.

The packs include a wide array of data, such as title data, local authority planning permissions, environmental assessments and energy efficiency ratings. With open data standards, this information is presented in a uniform, structured way that is easy for everyone in the property market to understand and use.

The model of collating information once that can then be re-used throughout the transaction significantly benefits from these standards. Once data is collected and transformed into a trusted data standard, its value multiplies, enabling sharing across the transaction's life cycle without the need for continuous onward transformation and reverification. This efficiency reduces errors and discrepancies, streamlining the process. A common infrastructure for data exchange makes these digital data packs easily discoverable and accessible for all parties in the transaction.

Interoperability, enhanced by open data standards, ensures that data from upfront data packs can be integrated across the wide variety of systems and platforms in the property industry. This seamless integration is crucial for collaborative and informed decision making among stakeholders, such as estate agents, solicitors, buyers and mortgage lenders. The efficiency gains are significant; with all necessary information in a single, digital format, the traditional process of gathering and verifying documents becomes much simpler. Moreover, this digital approach reduces the environmental impact by cutting down on physical documents.

Law Society transaction forms: the golden thread standard

The Law Society transaction forms, also known as TA forms, are a set of standardized documents used in property transactions in England and Wales.[12] Created by the Law Society of England and Wales, the forms aim to streamline the conveyancing process by ensuring that all relevant information is collected and shared consistently between parties involved in a property sale or purchase.

One of the most critical forms in this set is the TA6 Property Information Form. The TA6 form is completed by the seller of a property and provides the buyer with essential information about the property's condition, legal status and other relevant details.

Today, the TA6 form is typically completed early in the conveyancing process, shortly after the seller has accepted an offer from the buyer. The seller's conveyancer or solicitor will send the form to the seller to be filled out, along with other relevant forms depending on the type of property (e.g. TA7 for leasehold properties). The seller is responsible for completing the TA6 form to the best of their knowledge and belief. The form consists of several sections covering various aspects of the property, and it includes all questions relevant to the part A, B and C information defined by National Trading Standards.

The completed TA6 form is used by the buyer and their conveyancer to gain a better understanding of the property they are purchasing. The information provided helps the buyer make informed decisions and identify any potential issues that may need further investigation or negotiation. The buyer's conveyancer will review the TA6 form and may raise additional enquiries with the seller's conveyancer based on the information provided. This due diligence process helps to ensure that the buyer is aware of any risks or liabilities associated with the property before proceeding with the purchase.

Furthermore, the TA6 form serves as a legal record of the information provided by the seller. If any issues arise after the completion of the sale, the buyer may refer to the TA6 form to demonstrate that the seller had knowledge of the issue and failed to disclose it. This legal record can be crucial in resolving disputes or seeking compensation if necessary.

Though the Law Society transaction forms serve the conveyancing industry well, they were not designed for end-to-end property transactions. As a result, various other forms from other organizations and trade bodies that capture the same information are used for different parts of the transaction and by different people involved in the process. For example, estate agents often use the Property Information Questionnaire to capture property information from the seller. The issue with this approach is that the seller has to manually fill in this lengthy form and then later complete the TA6 form when requested by their conveyancer. This leads to the seller providing the same information to multiple parties in different formats, causing frustration and inefficiency.

The lack of a standardized, comprehensive form that covers all aspects of the property transaction can lead to information silos, inconsistencies and potential miscommunication between the various stakeholders involved. This fragmentation of data can slow down the process, increase the risk of errors and ultimately lead to a less satisfactory experience for the seller, the buyer and all professionals involved.

To address these challenges, there is a growing need for a more streamlined and integrated approach to property transaction forms. A single, standardized form that captures all the necessary information from the seller at the outset of the transaction and enables it to flow through the whole process would make for a more efficient and less frustrating transaction for all parties involved. This aligns with principle 7 of creating a golden thread: ensuring that information is clear, focuses on user needs and uses standard methods and terminology to allow for a consistent understanding.

A golden opportunity

Given that the Law Society transaction forms currently provide the standard method and terminology used by conveyancers to ensure consistency and understanding, there is a great opportunity for these forms to be the means of achieving standardization and consistency in the PTGT.

In 2024 the Law Society updated the TA forms to include all the material information required by National Trading Standards and allowed estate agents and other actors in the transaction to use their forms. This update presents a significant opportunity for creating a standardized property data pack at the start of the property transaction, using the Law Society transaction forms as the foundation.

By digitalizing the TA forms and integrating them into a comprehensive standardized property data pack, the data can flow seamlessly through the transaction life cycle, creating a PTGT of accurate, verified and immutable data in the reliable and trusted Law Society format. This digital transformation would allow for the accumulation of data throughout the process, with the complete dataset flowing to the Land Registry at the end of the transaction.

The benefits of using the data in the digital Law Society transaction forms to create the PTGT are numerous. For starters, it would reduce duplication of effort, as sellers would only need to provide information once, and this information would then be used throughout the transaction, saving time and reducing frustration. Furthermore, by using a standardized, comprehensive form, the risk of inconsistencies, errors and omissions would be minimized, leading to higher-quality data.

The digital flow of data would streamline the process by reducing manual data entry, minimizing delays and enabling faster completion of transactions. This more streamlined, efficient and transparent process would lead to a better experience for buyers, sellers and professionals involved in the transaction. Additionally, as the material information required by National Trading Standards is already incorporated into the TA forms, compliance with regulations would be more straightforward and easier to demonstrate. This would not only simplify the compliance process but also provide greater assurance to all parties involved that the necessary legal requirements are being met.

Finally, the seamless flow of data from the standardized property data pack to the Land Registry would facilitate a more efficient and accurate registration process. By having a complete, standardized dataset available at the end of the transaction, the Land Registry would be able to process the registration more quickly and with fewer errors, benefiting all parties involved.

Conclusion

The concept of a property transaction golden thread – built on the principles outlined in the Hackitt Report and the Building Regulations Advisory Committee's 'golden thread' report – offers a promising solution to the data trust issues currently plaguing the property market. By leveraging the existing Law Society transaction forms, and particularly form TA6, as the foundation for a standardized, digital data pack, the industry can create a more efficient, transparent and secure transaction process.

The digitalization of these forms, combined with the application of the ToIP framework, can help establish a secure and trusted digital environment that supports the principles of accuracy, security, accountability and accessibility. This digital transformation would enable the seamless flow of verified, immutable data throughout the transaction life cycle, reducing duplication of effort, improving data quality, enhancing transparency and ultimately leading to a better user experience for all parties involved.

Moreover, the adoption of open data standards and the integration of the standardized property data pack with the Land Registry would further streamline the process, facilitating compliance with regulations and ensuring a more efficient and accurate registration process.

As the UK moves closer to approving the Data Protection and Digital Information Bill and implementing the digital identity and attributes trust framework, the property market has a golden opportunity to create a property transaction golden thread. By doing so, the industry can address the long-standing issues of data trust and inefficiency, ultimately benefiting buyers, sellers and professionals alike, and ushering in a new era of transparency, reliability and efficiency in property transactions.

CHAPTER 8

Digital identity

> The Internet was built without an identity layer.
>
> — Kim Cameron, Chief Architect of Identity, Microsoft

If we think about the things we own that we might want to bring into the digital world, then our identity is arguably our most valuable asset. It is something we often take for granted, and we rarely worry about its loss. However, in the digital age, both its value and its vulnerability have become increasingly apparent. Therefore, when considering the digitalization of our physical lives, we must prioritize the safe, secure and controlled transfer of our identities, and this means our identities both as individuals and as businesses.

In this chapter I review the current state of identity in the online property market. I introduce the legislative and technological changes on the horizon that are set to revolutionize the way consumers and businesses manage and maintain control over their identities in the digital world.

The missing identity layer

An identity layer, in the simplest terms, is a means for each user to verify themselves online. Today, the internet allows anyone to access it anonymously, as it does not have a native identity layer. As a result, every business or platform on the internet has to build its own identity management system. If the internet had an identity layer, users would have a universal internet ID that they could use across all websites, similarly to how a passport (and appropriate visa) can give you access to any country.

In the early days of the internet, you had to physically sit at a designated terminal, often within a controlled environment such as a university or military base. Access was inherently secured by your physical presence and the permissions granted by the institution. In many ways, it was similar to accessing a secure facility today, where you need to present an ID card or pass a biometric check to use certain resources.

The introduction of personal computers in the 1980s meant the end of this physical security, but it also democratized access to the internet. When the World Wide Web came around in the early 1990s, it was not just a select few who could surf the web – it was anyone with a PC.

Back then, the web was mostly for reading: users consumed content from servers.[1] In this era, which would come to be known as Web 1.0, websites were like shop windows, allowing you to look but not touch. As a business, you did not need to know who was browsing unless they stepped in to buy something.

The evolution from a read-only web to an interactive one (discussed in chapter 3) happened gradually over the late 1990s and early 2000s. It was not until Dale Dougherty started to discuss 'Web 2.0' around 2004 that the concept of an interactive web really took hold.[2] This era introduced blogs, social media and wikis – platforms that empowered users to create and share content, forming vibrant online communities. It marked the dawn of usernames and passwords as the new gatekeepers to digital resources, a rudimentary attempt to add the missing identity layer.

This shift to a more social and interactive web was a boon for e-commerce. Platforms such as eBay and Amazon flourished, leveraging user reviews as a core aspect of the shopping experience. The role of user-generated content in marketing became undeniable, with testimonials and reviews enhancing trust and sales. Targeted advertising and direct customer interactions via social media further propelled e-commerce's growth.

The rise of secure online payment systems, alongside users' increasing comfort with online engagement, spurred an e-commerce explosion. This trend extended into the property market, with consumers beginning to handle aspects of property transactions online. Property portals such as Rightmove became popular for browsing listings, and user accounts for interacting with estate agents, conveyancers, brokers and lenders became commonplace, marking a significant shift in how property transactions are initiated and managed.

Figure 3. On the internet, nobody knows you're a dog.

But as e-commerce has grown, so have the risks around digital identities. Originating from a Peter Steiner cartoon published in the *New Yorker* in 1993, the phrase 'On the internet, nobody knows you're a dog' humorously highlights the internet's anonymity.[3] While users may have usernames and passwords, there is often no direct link between these digital credentials and their real-world identities. This absence of a reliable identity layer affects not just individuals but also businesses. For example, it is easy for fraudsters to create a domain name and email address impersonating reputable entities such as law firms and estate agents.

The missing identity layer hurts the property market

The shift towards digital property transactions has introduced significant challenges related to the verification and management of identities. In today's online property transaction, each participant often juggles multiple digital identities across various platforms, from property portals and agent websites to broker and lender portals, including platforms for 'know your customer' (KYC) procedures. This scenario potentially triples the number of identities involved, leading to a tangled web of siloed and disconnected digital personas.

This proliferation of unverified and fragmented digital identities not only complicates seamless communication and transactions among parties but also significantly undermines the integrity and efficiency of property dealings online. The absence of a unified digital identity layer in the property market therefore poses a considerable challenge.

Take the common scenario of Jo Bloggs, a consumer navigating the property market. As Jo engages with different parties involved in the property transaction – agents, brokers, lenders and conveyancers – she encounters a disjointed process of digital identity verification. Initially, Jo creates an account on a conveyancer's portal, undergoing basic email verification. She is then redirected to another platform for KYC verification, resulting in a verification document that is not linked back to her original account. This process, repeated across different platforms for agents, brokers and lenders, not only multiplies Jo's digital identities but also leaves them unverified and disconnected from each another.

As Jo interacts with additional platforms, each requiring its own set of verifications and accounts, the fragmentation becomes more and more complex. Instead of a single, verified digital identity, Jo ends up with multiple, siloed identities across the transaction chain. This not only complicates interactions but also introduces significant risks.

Firstly, the lack of a unified identity verification process makes Jo more vulnerable to identity theft and fraud, since malicious actors can exploit gaps in the system, compromising the security of Jo's personal information and the integrity of the transaction.[4]

Secondly, replicating personal information across multiple platforms without proper linkage or verification creates unnecessary data redundancy. This not only increases the risk of data breaches but also violates principles of data minimization and consumer control over personal information.[5] Jo's data is scattered across various systems, making it difficult for her to keep track of where her information is stored and how it is being used.

Thirdly, managing and verifying disjointed identities across platforms is time-consuming and resource-intensive for all parties involved. It slows down transactions, increases operational costs and leads to a poor user experience.[6] The inefficiencies resulting from fragmented digital identities can significantly hinder the smooth functioning of the property market.

Lastly, the absence of verified, unified digital identities undermines trust in digital property transactions. Consumers like Jo are left uncertain about the security of their information and the integrity of the transaction process.[7] This lack of trust can deter individuals from engaging in digital property transactions, hindering the growth and development of the market. To mitigate these issues, the property market urgently needs to adopt a more integrated and mature digital identity system. The government's recently proposed digital identity and attributes trust framework, discussed later in the chapter, has the potential to meet this need.[8]

By implementing a standardized and interoperable digital identity system, the property market can enable consumers like

Jo to have a single, verified digital identity that can be securely shared across different platforms and all the parties involved in the transaction. This would not only simplify the process for consumers but also reduce the risk of fraud, data breaches and operational inefficiencies.

Moreover, a mature digital identity system would allow for the secure and controlled sharing of personal information so that data is only shared with the necessary parties and for specific purposes. This would give consumers like Jo greater control over their personal information, enhancing privacy and building trust in the digital property market.

HM Land Registry is missing an identity layer

The lack of an account-based digital identity service at HM Land Registry represents a significant impediment to the digital property market and the digital economy. While individuals can manage their bank accounts, their social media profiles and even their tax returns online, the process of accessing property titles remains arduous and digitally outdated. This deficiency transcends mere inconvenience, exposing a profound gap in the property transaction framework that sets the Land Registry apart from nearly every other online service provider in the modern world.

In today's digital world, it is almost unheard of for a platform or organization to operate without an account and user verification service. From e-commerce giants like Amazon to government services like GOV.UK, the ability to create a secure, personalized account is a fundamental feature that enables users to access and manage their information efficiently. Even smaller online retailers and service providers prioritize account creation and identity verification as a means of enhancing user experience, security and data management.

However, the Land Registry's lack of a digital identity layer to securely associate property records with their legitimate owners and related parties is a critical capability gap that places it far behind the best practices of the digital world. This lack of an identity layer is not only inconvenient for users but also exposes the property transaction process to significant risks and inefficiencies.

Consider this scenario: the Land Registry recognizes hundreds of titles under the name 'John Reynolds', yet it cannot distinguish between individuals with that name. This essential identification task is outsourced to conveyancers. In contrast, if an individual named John Reynolds were to create an account on an e-commerce platform or the website of a government service, he would be required to provide unique identifying information, such as an email address, a phone number or even a government-issued ID. This information would be used to create a secure, personalized account that ensures John Reynolds can access and manage only his own data, without confusion or interference from other individuals with the same name.

The absence of a secure, account-based system also introduces a significant security risk, as the process of verifying property ownership and managing property-related transactions remains heavily reliant on outdated, paper-based methods. These manual processes, aside from being inefficient and error-prone, are also highly susceptible to fraud and identity theft. Dealing with these problems requires a significant headcount, and the Land Registry recently added 1500 case workers to their staff.[9]

Furthermore, the lack of a digital identity layer at the Land Registry hinders the development of a truly integrated, end-to-end digital property transaction process. Without the ability to securely link property records to verified user accounts, it becomes difficult to streamline the flow of information between different stakeholders in the property market, such as conveyancers, lenders and government agencies. This fragmentation of data and processes not only slows down transactions but also increases the risk of errors and inconsistencies.

Business identification

As we have seen, in today's digital market the verification of customer identities in high-value property transactions is a critical concern. However, an equally important aspect that often goes overlooked is the verification and continuous reverification of the identities of each business involved in a transaction.

Verifying businesses' identities goes beyond mere due diligence; it is about ensuring a secure environment in which transactions can safely occur. Despite the internet's vast capabilities, it does not inherently guarantee security – particularly in property transactions, for which the financial stakes are significant. The process of verifying businesses' identities is crucial not only for mitigating risks but also for maintaining the integrity of the entire transaction process.

The complexity of this challenge lies in the fact that verifying businesses involves more than just knowing who is on the other side of the transaction and confirming their legal status and adherence to applicable regulatory standards. It also requires establishing digital trust, which ensures that the two digital identities can connect and interact securely. This digital trust must be built on a foundation of robust identity verification processes, secure communication channels and transparent data sharing practices.

To achieve this level of trust, businesses must adopt a comprehensive approach to identity verification that encompasses both the initial verification of a business's identity and its continuous reverification throughout the transaction process. This ongoing verification is necessary to ensure that the business's identity remains valid and has not been compromised or altered in any way.

Moreover, establishing digital trust between businesses requires a standardized framework for identity verification and data sharing. This framework should define clear guidelines for how businesses can securely exchange information, verify each other's identities and establish trust online. By adhering to a common set of standards and best practices, businesses can create a more secure and efficient environment for property transactions.

Case study: Swift

The Swift network offers an exemplary model of how rigorous identity verification can fill the gap left by the internet's lack of a native identity layer. As described in chapter 5, Swift plays a pivotal role in global finance by facilitating secure, standardized and reliable communication between financial institutions worldwide. At the heart of Swift's operational success is a comprehensive identity verification system that instils in all its members a deep trust in the security of the network.[10]

When a financial institution seeks to join the Swift network, it embarks on a process that culminates in the issuance of a unique Swift bank identifier code (BIC). This alphanumeric sequence acts as a digital identifier, akin to a global passport for financial transactions on the Swift network. The issuance of the code follows a stringent process, beginning with an application that verifies the institution meets Swift's high operational and regulatory standards. A thorough due diligence review confirms the institution's legal, regulatory and reputational standing. Upon successful review, Swift assigns a BIC and registers it in a globally accessible directory.

Beyond just assigning a code, Swift issues digital certificates to institutions, providing cryptographic evidence of identity. These certificates are crucial for authenticating each Swift message, certifying its origin and ensuring it has not been altered in transit.

Swift's approach addresses the internet's inherent anonymity by ensuring the uniqueness of each BIC, guaranteeing that messages can be precisely traced back to their source. This level of authentication, paired with message encryption, safeguards information by ensuring it reaches only its intended recipient and gets there with its integrity intact. Moreover, the Swift Customer Security Programme mandates continuous compliance, reinforcing secure practices.

The benefits of Swift's identity verification framework are manifold. For one, it effectively solves the problem of the internet's missing identity layer by providing a secure, trusted method for financial institutions to verify each other's identities. This trust is vital, as it allows members to confidently engage with one another, knowing that the entities on the other side of a transaction are exactly who they claim to be. This assurance drastically reduces the risk of fraud and enhances the reliability of global financial communications.

Swift's identity verification framework sets a benchmark for digital identity security, demonstrating the significant advantages of having a secure, trusted identity layer.

Comparing the property market to Swift

There is a fundamental difference in how consumers and businesses in the property market identify themselves compared with those on the Swift network. In the property sector, the absence of a standardized identity framework leaves consumers and businesses navigating a fragmented and insecure online market. Parties are forced to depend on a mix of paperwork, external verification services and, frankly, blind trust when engaging with new partners or providers. This situation not only introduces delays but also heightens the risk of fraud and errors, making every transaction fraught with uncertainty.

This lack of a cohesive digital environment restricts the ability of businesses to fully automate and streamline their operations, leading to inefficiencies and missed opportunities. Transactions drag on, costs escalate and the potential competitive advantage of digital streamlining is lost.

What the property market desperately needs is a system in which the digital identities of consumers and businesses are as easily and securely verified as those within Swift. A few years ago, proposing such a transformation might have seemed overly optimistic. However, with the introduction of the Data Protection and Digital Information Bill and the accompanying UK digital identities and attributes trust framework, a cohesive digital identity framework for the property market is now within our grasp.

Digital identities and attributes trust framework

The introduction of the 2024 Data Protection and Digital Information Bill represents a decisive move by the UK government to reshape digital identity services, aligning them with the needs of the digital economy.[11] This bill signifies the government's commitment to transforming digital identity to make it more effective and user-centric. As Eric Schmidt stated in *The New Digital Age*: 'Identity will be the most valuable commodity for citizens in the future, and it will exist primarily online.'[12]

Back in 2014, in *Identity Is the New Money*, David Birch advocated for a National Entitlement Scheme:

> Suppose the vision for national identity shifted focus to certificates rather than cards or biometrics. As a participant, I could have a certificate on a national entitlement card, my bank card, or my mobile phone. These certificates would prove my entitlements (e.g. NHS access, office entry, shopping at Waitrose), emphasizing what I'm allowed to do rather than who I am.[13]

This was an early definition of the kind of framework that, ten years on, the upcoming Data Protection and Digital Information Bill is designed to introduce. This new legislation is set to revolutionize how digital identities are used and verified across the UK, especially in the digital property market.

The legislation will then open the way to creating a digital identity and attributes trust framework.[14] This framework is designed to return the control of digital identity to the users, enabling them to use their digital identities across multiple services and share their details effortlessly. To understand how the framework functions, we first need to understand a little more about the concept of identity.

Understanding identity and attributes

The World Economic Forum gives the following definitions of identity and identification.

- Identity: who a person or organization fundamentally is – a combination of attributes, beliefs, personal/organizational history and behaviour that together constitute a holistic definition of the individual or organizational self.

- Identification: the act of verifying identity; proving who people and organizations say they are.[15]

The digital identities and attributes trust framework states that

> a digital identity is a digital representation of a person acting as an individual or as a representative of an organization. It enables them to prove who they are during interactions and transactions. They can use it online or in person. Services and organizations that let users use secure digital identities can better trust that those users are who they say they are.[16]

Attributes are fundamental concepts within the digital identity and attributes trust framework. They are pieces of information that describe specific aspects of a person or an organization. These attributes can be combined to create a digital identity, but first, they must be 'bound' to the individual or entity.

> Attributes can represent characteristics possessed by a person or an organization, items or documents issued to them, or aspects related to devices or physical/digital documents. Examples of attributes include ones related to personal identity, such as name, address and date of birth; demographic attributes, such as number of children or age (e.g. whether someone is over 18); financial attributes, such as bank account number or number of employees; official identifiers, such as National Insurance number, NHS number or Companies House number; and qualifications or certifications, such as university degrees or professional accreditations.
>
> Attributes serve not only to create digital identities but also to verify eligibility or entitlement for specific actions. In some cases, this proof can be added to an existing digital identity, while in other cases the user's identity may not need to be known to complete an interaction or transaction.

The framework establishes a set of practices and processes that need to be followed when creating a digital identity. These include the requirement that 'all identity service providers … use the guidance on how to prove and verify someone's identity as a methodology to explain their product or service.' This refers to the guidance provided by Good Practice Guide 45, a document published by the UK government's National Technical Authority for Information Assurance, now part of the National Cyber Security Centre.[17] The document provides guidance on the process of verifying an individual's identity for the purposes of issuing digital credentials or granting access to systems and services.

For users, this translates into more streamlined, faster and safer digital interactions. They gain greater control over their personal information, with the framework incorporating privacy regulations that empower users to decide which personal details to use for their digital identity. Additionally, the framework emphasizes data minimization by ensuring that only necessary information is shared. For instance, when purchasing age-restricted products, only the user's age is required, and not the other details from their ID, minimizing the risk of identity fraud.

Oversight of the trust framework will initially be handled by a governing body set up by the Department for Digital, Culture, Media and Sport. As the market evolves, a more permanent structure will be established.

For organizations to participate, they must obtain certification against the framework's rules. This certification process involves an independent evaluation to ensure compliance with all regulations. Certified organizations are then awarded a trust mark, which enhances user confidence in their services. The framework encompasses various roles for organizations, each with specific rules to follow in adherence with the corresponding set of regulations.

- Identity service providers are responsible for confirming users' identities and can be either public or private entities. They offer a range of services, including identity verification, authentication and fraud checking, and they can share verified identities with other parties for service access or account creation.

- Attribute service providers manage specific user information and share attributes in line with user consent rules, ensuring the quality of the attributes they handle.

- Orchestration service providers are key to secure data sharing within the framework, assuming sub-roles such as identity brokers or hub service providers.

- Relying parties are entities that use services provided by others within the framework, mainly for identity verification or eligibility checks. They adhere to the framework's security standards, though they may not require certification unless offering additional services.

The government's investment of time and resources in creating the legislation, developing the framework and establishing the accreditation service underscores a strong political commitment to revolutionizing digital identity. This development marks a pivotal moment in the journey towards a more efficient, secure and user-friendly digital identity system.

Self-sovereign identity

One of the most exciting aspects of the digital identity and attributes trust framework is the implicit assumption that the UK is on a path to empowering consumers and citizens with self-sovereign identity (SSI). The language within the framework echoes the SSI movement, whose technical specifications are rooted in the World Wide Web Consortium's 'verifiable credentials' data model.[18]

This shift indicates that the property market is transitioning from a Web 2.0 identity model, characterized by centralized control, to a Web 3.0, decentralized, SSI model.

SSI is a transformative approach to managing digital identities, fundamentally shifting how personal information is controlled and shared. Unlike traditional models in which organizations store and control personal data, SSI empowers individuals to be the masters of their own identity information.[19]

One of the most compelling benefits of SSI is its emphasis on data minimization. In the context of property transactions, which currently require consumers to submit extensive personal information at multiple stages, SSI presents a significant improvement. This model enables individuals to provide only the necessary data required for a transaction, without the need to overshare or repeatedly submit the same details.

SSI addresses the inefficiencies and security concerns of the current process, in which sensitive information is often shared through unsecured means such as PDFs and emails. With SSI, verification occurs through digital credentials that are selectively disclosed. This means that individuals can prove their identity or qualifications without exposing any additional information. For example, if a transaction requires proof of address, an individual can share a credential verifying this fact alone, without revealing any unrelated personal details.

Moreover, SSI simplifies the process of identity verification in property transactions. Once an individual's identity is verified, they will possess a digital credential that can be re-used across different platforms and transactions. This both reduces the burden on the consumer of continuously providing documentation and streamlines the transaction process for all parties involved.

In essence, SSI represents a shift towards a more secure, efficient and consumer-friendly approach to identity verification. It eliminates the need for redundant data sharing, enhances privacy and improves the overall experience of engaging in property transactions.

Conclusion

The digital property market faces significant challenges due to the absence of an inherent digital identity layer, creating inefficiencies and security risks in property transactions. If we reflect on the internet's evolution and the Swift network's successful identity verification model, it is clear that adopting a similar framework within the property sector is essential. The introduction of legislative measures such as the Data Protection and Digital Information Bill and the digital identity and attributes trust framework, along with the potential of SSI, offers a roadmap towards a secure, efficient and user-centric property market.

To modernize the property market and ensure its operations align with the standards of the digital age, industry leaders and policymakers must unite to implement these technological and legislative advances. This collaboration will pave the way for a property market characterized by enhanced security, efficiency and trust, benefiting all stakeholders involved.

PART III

DIGITAL MONEY AND TITLES

In the following chapters I explore the potential of tokenized money and titles to enable a truly digital property market. At its core, the property market revolves around exchanging money for property ownership. While the market functioned effectively with physical money and deeds, the gradual dematerialization of both – coupled with the internet's missing identity layer – has introduced inefficiencies and disrupted seamless value exchange.

We are witnessing an exciting transformation in money and property titles. Money is being tokenized and going fully digital, while HM Land Registry continues to explore tokenized property titles. This profound transformation of money and titles into natively digital assets can potentially restore to the digital world the speed, ease, security and certainty of face-to-face transactions.

CHAPTER 9

Tokenization

> Software is eating the world, but blockchain could eat software. The implications of tokenization are profound, offering a new paradigm for asset ownership and exchange.
>
> — Marc Andreessen, co-founder of Andreessen Horowitz

Purchasing my first Ethereum token in 2017 opened up a new world of possibilities, capturing my fascination with the emerging concept of the 'internet of value'. I dedicated myself to learning about this new paradigm by reading, prototyping, attending meet-ups and essentially diving deep into the blockchain rabbit hole.

This chapter aims to provide an overview of tokens and their role in transferring real-world assets and identities into the digital world. By creating a digital environment in which your identity, money, contracts and counterparties coexist, you can connect, interact and transact seamlessly.

Value linked to physical tokens

The representation of value through tokens has a rich history dating back to ancient Mesopotamians around 3000 BCE, who used clay tokens to record transactions.[1] In 630 BCE the Lydian kingdom of Anatolia introduced a groundbreaking innovation by marking a piece of precious metal with a device resembling a signet ring. This innovation led to the creation of early coins and, eventually, to the *nomismata* used in the Byzantine Empire.

Tally sticks emerged as a straightforward and efficient method for documenting debts. Inscriptions on the sticks detailed financial obligations: they are the reason we know, for example, the debt that a bishop named Fulk Basset owed to King Henry III for the farm of Wycombe.[2] The tally stick system involved splitting a stick into two halves: the 'foil' for the debtor and the 'stock' for the creditor. The system also introduced the innovative concept of a tally stock holding value equivalent to the debt it represented, assuming trust in the debtor's ability to pay. This transformation turned tally sticks into a tradable form of money that could circulate freely among individuals.

The history of coinage in Britain began around 80 BCE with the arrival of Celtic tribes and saw a significant shift following the Roman invasion in 43 CE, which introduced Roman coinage.[3] Locally minted coins were rare, with the coins circulated in Britain typically being imported from Rome or its provincial mints.[4]

The Royal Mint's history dates back to 973 CE, when King Edgar was crowned the first king of a united England. Under Sir Isaac Newton's stewardship as master of the mint in the late seventeenth century, the Royal Mint saw substantial technical advances and improved standards that combated counterfeiting and ensured the reliability of British coinage.[5]

The Bank of England, established in 1694, complemented the Royal Mint's efforts by issuing banknotes. These banknotes evolved from paper promises of payment in coins to legal tender in their own right.

The establishment of the East India Company in the seventeenth century introduced share certificates, which represented a new form of value and ownership. These certificates served as precursors to modern stocks. The emergence of stock exchanges, such as the London Stock Exchange, played a critical role in facilitating financial activities.

Throughout this history, value has been closely linked to physical items such as gold, metals and paper, which have served as tangible symbols for trade and wealth. Similarly, ownership of assets such as property has traditionally been documented through physical deeds and titles. These material objects share the following characteristics.

- *Uniqueness.* Each item, whether a £50 note or a property deed, is unique in its individual features and history.

- *Verifiability.* Physical tokens of value and ownership documents can be authenticated, allowing their genuineness to be verified and confirmed.

- *Atomic transferability.* Tangible cash and deeds enable seamless, all-or-nothing transactions, in which either both the cash and the deed are exchanged or neither transaction takes place.

- *Possessability.* Property owners can physically hold their deeds as evidence of their rights, while buyers can possess their currency in a tangible form.

The evolution of value representation and ownership documentation has been a crucial aspect of economic development throughout history. From the earliest forms of tokens to the introduction of coins, banknotes and share certificates, the representation of value has undergone significant transformations. Similarly, the documentation of land ownership has evolved from early records such as the Domesday Book to modern property deeds and land registries.

The characteristics of uniqueness, verifiability, atomic transferability and possessability have been essential in establishing trust and facilitating transactions. These attributes have provided a solid foundation for the exchange of value and the transfer of ownership rights, enabling the growth and development of economies.

As we move into the digital age, the representation of value and ownership is undergoing another significant shift. The emergence of digital currencies, digital tokens and blockchain technology is challenging traditional notions of value and ownership. However, the fundamental principles of uniqueness, verifiability, atomic transferability and possessability remain relevant in the digital world.

The challenge now lies in translating these principles to digital representations of value and ownership, enabling them to be trusted, verified and exchanged with the same level of confidence as their physical counterparts. By leveraging technology and establishing robust frameworks for digital value and ownership, we can build on the rich history of value representation and ownership documentation and usher in a new era of economic growth and development.

Identity linked to physical things

Like value, identity in the physical world has been linked to physical things. Seals and signatures were the primary methods for authenticating identity in legal and commercial transactions before photo IDs became commonplace. A unique seal or handwritten signature provided a personal mark that could be compared against known samples for verification.

Birth certificates, marriage licences and government-issued identification cards and passports are among the most common physical documents still used to prove identity. These documents typically feature the individual's name, photo, signature and other identifying details that can be visually inspected and compared against the person presenting them.

The presence of witnesses has historically served to confirm the identities and actions of parties involved in a transaction. Witnesses can attest to the identity of the individuals and the legitimacy of their interactions, serving as a human form of authentication. In some cases, identity has been associated with physical tokens or objects other than documents, such as a key to a lock. The holder's ability to use the token indicates they are granted access by the owner, indirectly verifying their identity as someone trusted. Like other physical methods of proving identity, this relies on the principle that a person's identity can be confirmed through direct, material means.

The unintended consequences of dematerialization

The shift from physical to digital representations of money, deeds and identities in property transactions has resulted in unintended consequences, diminishing the utility of previously inherent features such as uniqueness, possessability and atomic transferability. The dematerialization of digital identities exacerbates this issue, as they have become fragmented and duplicated across various platforms and systems (see chapter 8). Below, I examine the impact of these changes on property transactions.

- *Uniqueness.* In the physical world the uniqueness of an item is established by the nature of the transaction: when you give someone money, they posses it while you do not. In the digital world, however, sending £50 from one computer to another does not guarantee that it has not been forwarded to multiple recipients. This issue is known as the 'double-spend' problem, and solving it means ensuring that no one can sell the same item to multiple buyers or use the same money for different purchases. To address the double-spend problem in the digital domain, we rely on trusted intermediaries such as banks for monetary transactions and HM Land Registry for property titles. These organizations facilitate secure online value transfers and prevent double-spending.

- *Possessability.* In the physical world you have tangible control over your money, deeds and identity. In contrast, the digital world transfers this control to third-party organizations. In the case of money, you entrust your digital identity and funds to the bank, which grants you access and management rights through an account. Although the bank holds your money, you maintain a sense of possession and control. However, when it comes to property deeds, you submit them to HM Land Registry without an equivalent identity service. Consequently, you cannot create an account or maintain a tangible sense of possession in the digital world, as you lack direct access to and control over your deeds.

- *Atomic transferability.* The real-time, seamless physical exchange of deeds and money is a feature of atomic transferability. However, in the digital world, an atomic transfer of money and deeds is not feasible. The transaction process (detailed in chapter 14) is affected by the dematerialization of money, deeds and identity, which cannot be exchanged or verified simultaneously and automatically.

The dematerialization of value and identity in the digital world has led to their fragmentation and duplication across various platforms and systems. This has resulted in a loss of uniqueness, possessability and atomic transferability, which were inherent features of physical representations. The reliance on trusted intermediaries, such as banks and HM Land Registry, has become necessary to address issues such as the double-spend problem and to facilitate secure online value transfers.

However, this reliance on intermediaries has also led to a loss of direct control and possession over one's money, deeds and identity. While banks offer account systems, the lack of an equivalent identity service at HM Land Registry means that individuals cannot maintain a tangible sense of possession or direct access to their deeds in the digital world.

Furthermore, the loss of atomic transferability in property transactions has meant that the seamless, real-time exchange of deeds and money practised in the physical world cannot be easily replicated in the digital world, leading to a more complex and fragmented transaction process.

To address these unintended consequences, it is crucial to develop digital systems and frameworks that can effectively replicate the unique features of physical representations. This may involve the use of emerging technological capabilities such as tokenization and smart contracts, which can help to establish uniqueness, possessability and atomic transferability in the digital world.

Managing digital tokens

A good introduction to the basics of tokens can be found in the 'Blueprint for the future monetary system' from the Bank for International Settlements.[6] In summary, the paper explains that while account managers in traditional ledger systems are responsible for maintaining and updating accurate ownership records, tokenized systems treat money or assets as 'executable objects' maintained on programmable platforms. These objects can be transferred through the execution of programming instructions issued by system participants without the need for an account manager's intervention.

While tokenization does not completely eliminate the role of intermediaries, it does change the nature of their role. In a tokenized environment, the operator serves as a trusted intermediary in a governance role, acting as the curator of the rule book rather than a bookkeeper who records individual transactions on behalf of account holders.

The claims traded on programmable platforms are known as tokens. Unlike simple digital entries in a database, tokens integrate the records of the underlying asset typically found in a traditional database with the rules and logic governing the

Token

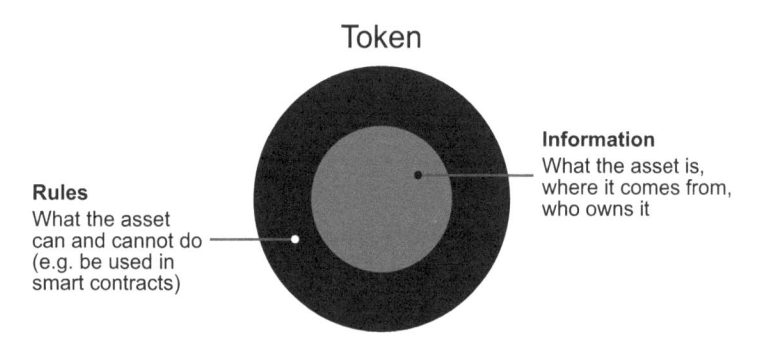

Rules
What the asset can and cannot do (e.g. be used in smart contracts)

Information
What the asset is, where it comes from, who owns it

Figure 4. Tokens both define assets and specify what can be done with them.

transfer process for that asset. This means that while traditional systems usually have common rules for updating asset ownership, tokens can be customized to meet specific user or regulatory requirements that apply to individual assets.

Tokenization: a solution to dematerialization

Tokenization in the digital world offers a promising solution to the challenges posed by dematerialization, as it specifically addresses the core issues of uniqueness, possessability and atomic transferability that are essential for transactions involving money, deeds and identities.

Tokenized money, or digital currency, such as bitcoin, uses blockchain technology to restore the attribute of uniqueness.[7] Each digital token represents a unique, non-duplicable unit of value, which effectively solves the double-spend problem without the need for traditional intermediaries. These tokens are cryptographic proofs that cannot be replicated, ensuring that when a digital currency is transferred, the sender no longer possesses it. This restores to digital transactions the confidence that was once exclusive to physical money exchanges.

Similarly, tokenizing property deeds ensures that every property has a unique digital representation. By issuing tokenized titles on a digital asset platform, each token becomes a digital stand-in for the physical deed, controlled exclusively by the token's owner. This means that the owner retains a digital version of possessability, as the entry on the digital asset platform can be transferred only with the owner's consent, typically through secure cryptographic keys. The entry is a single, immutable record, providing a transparent history of ownership and transactions associated with the property.

The programmability of digital tokens is another key utility. The Ethereum platform introduced the concept of smart contracts, or programmable objects.[8] This enabled tokens to include embedded logic and rules in the form of a self-executing contract. As a result, transactions and other actions can be automatically executed when certain conditions are met, without the need for intermediaries. This capability opens up endless possibilities for automating processes, creating self-executing agreements and developing complex decentralized applications that can serve a wide array of purposes, from finance to logistics and beyond.

Tokenization therefore brings back the concept of atomic transferability to digital transactions. Smart contracts – as self-executing contracts with the terms of the agreement written directly into code – can be used to facilitate the simultaneous exchange of tokenized money, deeds and identity verifications. This is covered in detail in chapter 14.

Tokenization stands as a transformative approach to reinstating the trust and utility that dematerialization removed. By leveraging the inherent benefits of blockchain technology, such as decentralization, immutability and cryptography, tokenization ensures that money, deeds and identities retain their fundamental attributes. This innovation not only mirrors the assurances of the physical world but also enhances them, providing a more secure, efficient and transparent system for property transactions in the digital age.

The token taxonomy

We can better understand tokens by looking at a well-established and broadly accepted taxonomy that classifies them based on their functions and underlying value propositions. This taxonomy aids not only in comprehension but also in discerning the various applications and legal considerations associated with each token type. For a full and in-depth look at token taxonomy, readers can refer to the work of the Token Taxonomy Initiative.[9]

One important distinction is between fungible and non-fungible tokens. Fungible tokens offer a new way to think about currency and asset shares in a decentralized context, enabling seamless transactions and access to capital. Characterized by their interchangeability, these tokens can be exchanged on a one-to-one basis with others of the same type, much like traditional currency. This attribute makes fungible tokens particularly suitable for use as digital currency, serving as a medium of exchange in the digital space. Moreover, they can represent shares in assets, allowing for fractional ownership and democratizing access to investment opportunities. Bitcoin and Ethereum's Ether are examples of fungible tokens. Their widespread acceptance and their use case as both a store of value and a means of transaction underscore the adaptability and potential of fungible tokens in the digital economy.

Non-fungible tokens (NFTs) introduce the concept of uniqueness and indivisibility to digital assets. Unlike their fungible counterparts, NFTs cannot be exchanged on a one-to-one basis, as each token represents something distinct. This uniqueness is particularly valuable in representing ownership of digital or real-world assets that are one-of-a-kind, such as artworks, collectibles or – as I will discuss later – property titles.

Tokens can also be classified based on their functions. Currency tokens, such as bitcoin, are designed primarily as a medium of exchange, aiming to function as digital money – though they are only rarely used in this way. Utility tokens, on the other hand, are not meant to be investments; rather, they are akin to digital coupons that grant holders certain rights or privileges, such as using a blockchain-based application or participating in a decentralized network.

Security tokens bridge the gap between traditional finance and the blockchain world by offering tokenized shares of assets such as property assets, stocks or bonds. Commodity tokens symbolize ownership of or entitlement to a physical or digital commodity, enabling commodities such as gold, oil or even digital art to be traded and owned in fractional amounts on the blockchain.

Lastly, governance tokens give holders the power to influence decisions within a platform or organization, and they serve as a cornerstone of decentralized autonomous organizations (DAOs) and other decentralized platforms.

Token-based versus account-based systems

I know from my own experience that exploring the differences between token-based and account-based systems can significantly deepen one's understanding of how digital assets and currencies function within blockchain and traditional financial systems. An excellent discussion of these differences is set out by Tony McLaughlin in his paper on 'The regulated internet of value'.[10]

In token-based systems the essence of ownership is encapsulated by the possession of tokens. These tokens – which can represent anything from a unit of cryptocurrency to ownership rights in digital or physical assets – are stored in digital wallets. Transactions with tokens are straightforward: if I want to transfer a token to someone else, I simply send it from my wallet to theirs, mirroring the way physical currency is exchanged.

Account-based systems, on the other hand, are the familiar structure of traditional bank accounts. In these systems, ownership and transactions are not tied to the possession of tokens but are recorded as entries in a centralized ledger or database. Each account has a balance, and transactions are executed by adjusting these balances under the ledger's oversight, which is typically maintained by a trusted financial institution or entity. The reliance on a central authority introduces potential points of failure, including fraud, censorship and the risk of a single point of control. Token-based systems, with their emphasis on decentralization, aim to mitigate these risks by distributing the ledger across countless nodes, making it virtually tamper-proof and resistant to central control.

The distinction between token-based and account-based systems is not just a technical one; it has profound implications for how we conceptualize and interact with digital assets and currencies. Token-based systems, with their emphasis on decentralization and individual ownership, represent a paradigm shift in how we think about value and exchange in the digital age. They offer the potential for greater financial inclusion, transparency and resilience, as well as the ability to tokenize a wide range of assets and create new models of ownership and investment.

However, the transition to token-based systems is not without its challenges. Regulatory frameworks, which have been built around account-based systems, will need to adapt to accommodate the unique characteristics of tokens and the decentralized networks they operate on. Issues of privacy, security and scalability will also need to be addressed in order to ensure that token-based systems function effectively.

Additionally, the unregulated, anonymous nature of cryptocurrencies has resulted in their widespread use by criminals for activities such as money laundering. For example, one recent paper concluded that a quarter of bitcoin users are involved in illegal activity, and that the $76 billion in illicit payments involving bitcoin represented 46% of the currency's total transactions.[11] This illicit use of cryptocurrencies has resulted in the creation of regulated digital currencies, which I look at in detail in the next chapter.

Conclusion

In this chapter I have explored the historical relationship between value, identity and physical tokens, as well as the challenges and opportunities presented by their digital counterparts. From ancient clay tokens to modern banknotes and share certificates, physical representations of value and identity have long served as tangible symbols of wealth, ownership and trust. However, the shift towards digital representations has resulted in unintended consequences that compromise these established features.

Tokenization has emerged as a promising solution to these challenges, offering a way to restore and even enhance the utility of uniqueness, possessability and atomic transferability in the digital world. Tokenized money, deeds and identities can provide secure, efficient and transparent transactions through the use of blockchain technology, addressing the shortcomings of dematerialization.

The classification of tokens as fungible or non-fungible and in terms of their various functions allows for a more comprehensive understanding of their potential applications and legal implications. Furthermore, the comparison between token-based and account-based systems highlights the decentralized nature of the former, which aims to mitigate risks associated with central authorities. Tokenization will play a crucial role in shaping the future of property transactions, finance and identity management, a topic I pick up again later when discussing the propertyverse in chapter 18.

CHAPTER 10

Digital money

> Today, the monetary system stands at the cusp of another major leap. Following dematerialisation and digitalisation, the key development is tokenisation – the process of representing claims digitally on a programmable platform. This can be seen as the next logical step in digital recordkeeping and asset transfer.
>
> — Bank for International Settlements, 2014[1]

At the core of a property transaction is the exchange of money for property. However, in the current market, money remains isolated within online banking systems, lacking the portability and versatility of cash and the capacity for immediate transactions.

In this chapter I explore the emerging world of tokenized digital money. The future promises truly digital currency that can be present in various digital environments, be programmed to perform specific actions based on certain events, and be more portable, transferable and usable across different platforms and environments.

I aim to provide you with a solid understanding of what tokenized, digitally native money is. I will also share insights into the key initiatives that the Bank of England, commercial banks and Mastercard are undertaking to bring this new form of digital currency into existence, setting the scene for the innovative 24/7 property transactions explored in the discussion of digital completion in chapter 14.

What is money?

There are two forms of money in the economy.

- Central bank: money that is a liability of a central bank. It is available to the public in the form of cash. It is also available to commercial banks in the form of central bank reserves.

- Private: money that mainly takes the form of deposits in commercial banks – that is, claims on commercial banks held by the public. This 'commercial bank money' is created when commercial banks make loans to households and companies.

Money has three functions: it acts as a medium of exchange, a store of value and a unit of account. Bank of England money establishes and maintains sterling as the unit of account for virtually all transactions in the UK economy and, in doing so, anchors the monetary system.

A more useful form of money

Ever since buying my first cryptocurrency in 2017, the concept of new forms of money has fascinated me. The notion that money could be programmable and operate based on specific circumstances or events seemed far more dynamic than the static nature of traditional bank-account balances.

Having cryptocurrency on a USB stick felt more tangible and personal than seeing digits in a bank account. Holding a £50 note, which you can freely transfer, gives you a sense of complete control, and curiously, the cryptocurrency on my desk felt similarly tangible, emphasizing my responsibility for its safety – much like cash.

However, cryptocurrency's divergence from traditional cash included significant volatility and a lack of regulation. Its value could wildly fluctuate daily. I recall purchasing cryptocurrency

DIGITAL MONEY 135

that either did not arrive or became virtually worthless. This lack of market regulation left me, a retail user, exposed to scams and market volatility. Unbeknownst to me at the time, the next seven years would see the emergence of regulated forms of money that offered the utility and advantages of programmable cryptocurrencies without their instability. This chapter explores the rise of regulated tokenized money.

What is tokenized money?

Andrew Bailey, the governor of the Bank of England, provided an insightful perspective on tokenized money in his 2023 speech 'New prospects for money'. He described enhanced digital (i.e. tokenized) money as essentially a unit of money that comes with the added capability of executing a wide range of actions. These actions, often part of smart contracts, can range from simple to highly complex. Importantly, Bailey highlighted that while the functionality and utility of money are expanded through tokenization, the intrinsic nature or 'singleness' of money remains unchanged. It is not the fundamental essence of money that alters, but rather what can be accomplished with it through tokenization.[2]

Building on Bailey's insights, I am going to define tokenized money as 'a digitally unique, executable object that is immutable, ownable and transferable, representing a unit of money'.

This definition emphasizes several key attributes of tokenized money: it is immutable, ensuring that once a transaction is recorded, it cannot be altered or erased; it is ownable, meaning that ownership rights are clearly defined and can be securely transferred; and it is transferable, allowing for a seamless exchange between parties. Each unit represents a distinct value, embodying the characteristics of traditional money in a digital form that uses blockchain technology for enhanced security, efficiency and programmability.

The internet of value

Ripple, developed by Ripple Labs, is a global system for real-time settlement, currency exchange and remittance. Launched in 2012 with the tag line 'The internet of value', it provided a distributed open-source protocol that supported various tokens, including fiat currencies and cryptocurrencies.[3] Ripple's main cryptocurrency, XRP, was designed for quick, secure and low-cost international transactions.

Ripple's internet of value has faced significant challenges around compliance with financial regulations. In December 2020 the US Securities and Exchange Commission sued Ripple Labs and its top executives, alleging that XRP tokens were sold as unregistered securities. This lawsuit centred on whether XRP, created and distributed by Ripple in a centralized manner, should be classified as a security under US law.

Central bank digital currency

A 2014 Bank of England video on YouTube, titled 'The economics of digital currencies', features an interview between two bank representatives in an east London coffee shop with a bitcoin ATM. The interview highlights how bitcoin was created independently of any central bank and discusses how the ATM facilitates the exchange of bitcoin for banknotes.[4]

This video was part of the bank's autumn bulletin, and it was supported by a paper titled 'Innovations in payment technologies and the emergence of digital currencies'. While the paper does not explicitly mention central bank digital currencies (CBDCs), it observes:

> The majority of financial assets, including loans, bonds, stocks, and derivatives, are now primarily electronic, indicating that the financial system is essentially a series of digital records. These records are held in a tiered structure but might be replaced in the future by various distributed systems.[5]

Fast forward to March 2016, and Ben Broadbent, in a speech at the London School of Economics, emphasized the significance of innovations in digital currencies and pondered the economic implications of central banks introducing their own digital currencies. The speech highlighted potential impacts on commercial bank deposits and credit creation.[6]

Then, in March 2018, Bank of England governor Mark Carney's discussion of the future of money at the inaugural Scottish Economics Conference similarly focused on CBDCs and their potential to transform the economy.[7]

A pivotal moment occurred in March 2020 when the Bank of England released a discussion paper on CBDCs, proposing an electronic form of central bank money for payments by households and businesses.[8]

Following this discussion paper, 2023 saw the final report of Project Rosalind, an experiment by the Bank of England and the Bank for International Settlements Innovation Hub to develop and test a prototype CBDC infrastructure, exploring over 30 use cases.[9]

The digital pound consultation

In January 2024 the Bank of England and HM Treasury responded to the results of a public consultation on the idea of a 'digital pound', which had received more than 50,000 responses.[10] The feedback underscored ongoing concerns over the potential introduction of a digital pound. In response, the Bank of England and HM Treasury expressed their commitment to addressing these concerns by proposing a series of measures that would regulate the digital pound, should it be implemented.

- *Legislative measures.* The government has pledged that the digital pound will not be launched without primary legislation being passed by both houses of parliament. This step ensures that the introduction of the digital pound would be fully supported by a legal framework.

- *Privacy protections.* Privacy has been identified as a fundamental aspect of the digital pound's design. Specifically, the personal data of users would not be accessible to the Bank of England or the government. Legislation for the digital pound would enshrine the privacy rights of users. The bank is exploring technological solutions to ensure that it cannot access personal data via its core infrastructure. Also, the digital pound would not be programmed by the bank or the government, with legislation reinforcing this principle.

- *Access to cash.* Despite the potential introduction of the digital pound, the government has enacted legislation to protect access to physical cash, ensuring its continued availability.

While no definitive decision has been made on a CBDC, there is a growing belief that its introduction in the UK is inevitable, and will likely be sooner rather than later, given the recent pace of development.

Facebook's Libra coin

In June 2019 Facebook released a white paper proposing the Libra payment system. The proposal initially garnered mild interest but quickly became a significant discussion point among commercial banks. Prior digital currencies such as bitcoin and Ethereum, despite their public appeal and substantial investments, were largely seen as secondary to sovereign currencies, and as not posing a substantial threat. Libra marked a departure from this perception.

Facebook's extensive user base represented a significant market presence. Its array of services – including Instagram, WhatsApp and Messenger – saw 2.1 billion daily and 2.7 billion monthly users, out of a world population of 7.7 billion. This reach, combined with its $60 billion revenue, had positioned Facebook as a potential challenger to commercial banks.

This development led to a noticeable acceleration in the move towards digital or tokenized commercial bank money and hastened progress on CBDCs. But as well as undeniably spurring on the sovereign banking sector's exploration of digital money options, Facebook's venture into digital currency attracted considerable regulatory attention. The Libra coin – later rebranded as Diem – faced significant scrutiny and regulatory hurdles. Ultimately, this led to the project's discontinuation, as stated in a press release on its website in January 2022:

> We are gratified that the ... Report on Stablecoins issued by the President's Working Group on Financial Markets validated many of Diem's core design features. Those features address not only the risks related to the issuance of a stablecoin, but also the risks associated with transferring stablecoins between parties.
>
> Despite giving us positive substantive feedback on the design of the network, it nevertheless became clear from our dialogue with federal regulators that the project could not move ahead. As a result, the best path forward was to sell the Diem Group's assets, as we have done today to Silvergate.[11]

This statement encapsulates the complexities and challenges that are faced when introducing a new stablecoin into the financial market, and particularly those around regulatory compliance and market impact.

The regulated internet of value

Within one year of the Libra coin white paper, in June 2021, Tony McLaughlin – an expert in Emerging Payments and Business Development at Citi's Treasury and Trade Solutions wing – published a paper on Citi's website titled 'The regulated internet of value'.[12]

The paper assesses the transformative potential of distributed ledger technology (DLT) for the future of digital finance. It outlines a vision in which the tokenization of digital value – particularly in the form of regulated liabilities such as central bank money, commercial bank money and electronic money (e-money) – could significantly enhance the efficiency, security and accessibility of financial systems globally. McLaughlin articulates a compelling argument for a shift from traditional account-based transactions to a token-based digital economy facilitated by DLT.

The paper begins with the premise that DLT offers a superior framework for representing and transacting digital value (the 'tokenization thesis'), positing that tokenized assets could create 'always on', resilient and programmable financial networks.

The paper distinguishes between various forms of money and highlights the potential for a future in which digital currencies are primarily represented by stablecoins and CBDCs. However, McLaughlin suggests a third, more inclusive path that encompasses the tokenization of all regulated liabilities.

A significant portion of the text is dedicated to exploring the ongoing transition towards digital money, categorizing current digital financial instruments into central bank money, commercial bank money and e-money. The discussion points to both the limitations and the potential of these forms in a digital context, and it emphasizes that they must evolve to meet the demands of a modern economy.

McLaughlin argues that DLT can offer unprecedented benefits including continuous operation, single sources of truth, programmability, instant settlement and the ability to represent a wide array of assets on a single ledger. He calls for a regulatory and technological shift to support the widespread adoption of tokenized regulated liabilities. This involves pivoting current explorations into CBDCs and undertaking multi-bank efforts to tokenize commercial bank money, alongside the potential regulation of stablecoins.

The ultimate vision presented is a global, DLT-based financial infrastructure that seamlessly integrates regulated liabilities and assets. This network would not only streamline transactions but also incorporate a range of financial instruments into a unified, programmable platform.

Regulated Liability Network

In November 2022 the 'Regulated Liability Network: digital sovereign currency' white paper was published, with joint authorship by Swift, BNY Mellon Treasury Services, HSBC Global Payments Solutions, Lloyds Banking Group, OCBC, ANZ, Wells Fargo, US Bank and Trust Financial Corporation.[13]

The Regulated Liability Network can be seen as a platform for innovation, similar to earlier schemes for new platforms such as CHAPS (the Clearing House Automated Payment System). Through a common application programming interface (API), the Regulated Liability Network gives businesses and software platforms access to both commercial bank and central bank money. You can think of it as similar to open banking – the key difference being that it not only gives access to money but also adds new functions and utility to money, making it programmable. This programmability allows money to be locked and moved as part of business workflows or on the basis of given events.

The 2022 paper builds on all the key themes outlined by McLaughlin in 2021. In line with that earlier paper, it calls for a broader perspective on digital currency, beyond just central bank liabilities. Its emphasis is on viewing regulated liabilities – which include central bank money, commercial bank money and e-money – as a unified entity. The paper proposes shifting from a CBDC focus to a more encompassing approach focused on digital sovereign currency.

A key point made is the technological neutrality of legal instruments. The paper highlights that the legal validity of financial instruments does not change with the technology used to record them. For instance, whether liabilities are recorded in paper ledgers, in databases or on a DLT platform, their legal standing remains constant. This concept underscores the potential of updating the national currency system within existing legal frameworks by using advanced technologies such as DLT.

The paper also discusses the challenges posed by the emergence of unregulated digital currencies, such as bitcoin and stablecoins, and the potential risks they bring to consumer protection, financial stability and the prevention of financial crime. The growth of these unregulated networks could potentially weaken the sovereign currency system. To counter this, the paper suggests bringing these novel digital currencies inside the regulated perimeter and enhancing traditional payment systems (something which the Financial Conduct Authority has now published plans for).

Overall, the paper emphasizes the importance of adapting the sovereign currency system to modern technologies, ensuring regulatory compliance and preserving the integral role of the nation-state in the monetary system. Peter Left, head of prudential liquidity management at Lloyds Banking Group, gave this summary of the changes envisioned by the paper:

> RLN is an exciting foundation for the industry to make regulated commercial bank money interoperable to an extent we have never been able to achieve before. Commercial bank money smart contracts could be designed to be smoothly interoperable across commercial banks and central banks. Having made regulated money interoperable to this extent, we can then open innovation opportunities for smarter, more competitive payments.

The UK is on the path to tokenized regulated liabilities

After the publication of the Regulated Liability Network white paper, UK Finance, in collaboration with EY and a group of members and stakeholders, embarked on an exploratory journey into the concept of a Regulated Liability Network. The goal of this initial 'discovery phase' was to uncover viable use cases for a proof of concept within the UK, tackling key queries to ascertain how the organizations should engage in the forthcoming 'experimentation phase', which would involve the development, creation and execution of one or more proofs of concept.[14]

The discovery phase was fruitful, paving the way for further development. The initiative has now progressed to the experimentation phase, with a definitive aim to launch live tokenized deposits by 2025.

Mastercard MultiToken Network

In July 2023 Mastercard published its MultiToken Network (MTN) white paper. In many respects, the MTN aims to be a version of the type of network outlined in the Regulated Liability Network white paper: a regulated network that enables the issuance of tokens that represent a claim on sovereign money, in this instance commercial bank money.[15]

The essence of the MTN is to transform a portion of bank deposits into digital tokens. These are not just any tokens; they are designed to directly represent bank deposits, offering a level of reliability comparable to the funds in your bank account.

The MTN looks to innovate on how businesses and consumers interact with the financial and digital asset markets. By converting regular bank deposits into tokenized deposits, businesses can engage in transactions that directly utilize customers' bank funds, combining the trustworthiness and stability of conventional banking with the agility and innovation of digital currencies.

The technical achievement of creating tokenized deposits lies in translating part of a bank's balance into a blockchain-compatible digital format. These tokens retain all the functionalities of blockchain technology, including interoperability with other tokens and the ability to partake in smart contracts within the MTN.

In the summer of 2023 Mastercard selected Coadjute, out of more than 140 global applicants, as one of the few businesses to develop a prototype application on the Mastercard MTN. I led this project for Coadjute, which involved settling a mortgage transaction using commercial bank tokenized deposits. The prototype solution was showcased to tier-one mortgage lenders in the UK.[16]

Legal and regulatory framework

In the Regulated Liability Network white paper, the legal and regulatory framework surrounding tokenized deposits is clearly outlined. If we reflect on the evolution of money, from paper ledgers in the nineteenth century to computer databases in the twentieth century, it is evident that while the methods of recording money have transformed, the legal definition and essence of money have remained unchanged.

The concept of a legal instrument, such as a bank's commitment to its customers, retains its significance regardless of the medium through which it is recorded. This principle holds true whether the instrument is documented on paper, stored in a digital database or inscribed on any other material. The transition from individual bank databases to a shared ledger system does not alter the fundamental legal nature of these instruments. Moving to shared ledgers represents a technological advance without affecting the legal and regulatory framework that governs money.

This perspective emphasizes that the essence of money transcends its physical or digital representation. It is fundamentally a legal agreement and a bond of trust between a bank and its customers. This bond remains firm whether the documentation of financial transactions is centralized or decentralized.

The approach of the Regulated Liability Network white paper in differentiating the private law aspects of digital assets from their regulatory characterization underscores a critical point: the adoption of new technology to facilitate existing regulated activities often requires minimal or no changes to private law or regulation. As Michael Voisin, a partner at Linklaters, notes:

> Although many (but by no means all) crypto and DeFi [decentralized finance] tokens and activities are designed or structured without reference to legal or regulatory considerations, the private law has adapted to apply conventional legal principles to such tokens and activities while regulators and policymakers separately adapt and apply regulation to them. By contrast, very little adaptation of private law, and often no adaptation of regulation, is required where new technology is deployed to deliver existing regulated activities. As a rule, there is generally no difference in the legal characterization of a deposit at a bank, whether it is recorded in a physical ledger, in an on-site hard drive, in the cloud, or on a distributed ledger.[17]

This understanding is crucial in navigating the legal and regulatory environment of tokenized deposits, and it underscores the adaptability of existing legal frameworks to technological innovations in the financial sector.

Blueprint for the future monetary system

The Bank for International Settlements (BIS) aims to support central banks in their pursuit of monetary and financial stability through international cooperation, and it acts as a bank for these central banks. The BIS has established Innovation Hubs globally, including a centre in London working with the Bank of England (more on this in chapter 14).

In June 2023, as part of its annual economic report, the BIS released a significant paper titled 'Blueprint for the future monetary system'.[18] The paper expands on the core theme of the Regulated Liability Network, articulating and endorsing the benefits of a 'unified ledger':

> This [paper] presents a blueprint for a future monetary system that utilizes the potential of tokenization to enhance the old and enable the new. The key components of this blueprint are Central Bank Digital Currencies (CBDCs), tokenized deposits, and other tokenized claims on financial and real assets. The blueprint envisions these elements converging in a novel type of Financial Market Infrastructure (FMI) – a 'unified ledger'.

If the Regulated Liability Network white paper was a proposal from commercial banks to central banks for collaboration on a shared infrastructure for sovereign money, then the 'Blueprint for the future monetary system' can be seen as the formal response, expressing acceptance of the proposal.

New prospects for money

This endorsement from central banks as to the necessity for advances in both commercial bank and central bank money was reiterated by Andrew Bailey in July 2023 in his speech on 'New prospects for money': 'We want to encourage more thinking and action in the world of enhanced digital bank deposits – sometimes called tokenized deposits. So, yes, this is a call to action, particularly to banks – don't leave central banks as the only show in town.'[19]

So it is pretty clear that there will be both commercial bank and central bank digital currencies. As David Birch puts it:

> No matter how beneficial it might be to society, CBDC has two well-understood problems… The first is that it would

disintermediate banks, and the second is that it might destroy banks (because bank deposits would be traded for zero-risk central bank money).[20]

Conclusion

In this chapter we have explored the concept of tokenized digital money and its potential to revolutionize financial services. We have seen how the limitations of traditional online banking systems and the volatility of cryptocurrencies have paved the way for the development of regulated tokenized money.

Thanks to the insights of Andrew Bailey and the initiatives of the Bank of England, commercial banks and Mastercard, we are witnessing the emergence of a new form of digital currency that combines the benefits of programmable cryptocurrencies with the stability and regulation of traditional money.

The journey from the early days of Ripple's 'internet of value' to the Bank of England's exploration of CBDCs and the development of Mastercard's MTN illustrates the rapid evolution of tokenized money. The future of money is becoming increasingly digital, portable, transferable and versatile, promising to transform the way we conduct transactions in the digital world. As we move forward, the legal and regulatory frameworks surrounding tokenized deposits will continue to adapt, ensuring the protection of consumers and the stability of the financial system.

The advent of tokenized digital money represents a significant step in the transformation of physical entities into digital ones. It offers a more useful, regulated and secure form of money that can be seamlessly integrated into various digital environments, opening up the possibility of real-world financial transactions within the emerging propertyverse.

CHAPTER 11

Digital titles

> We will continue to explore the potential benefit of emerging technologies such as tokenisation (the creation of a blockchain-based, digital representation of a real-world asset such as property), fractional ownership and other alternative models.
>
> — HM Land Registry Strategy 2022+[1]

Historically, ownership of property was represented through physical deeds, and today it is represented by the land register. However unlike money, which you can physically hold in the form of cash, and easily act on (by transferring it via online banking, for example), there is currently no way to easily carry out an action on your deed held by HM Land Registry.

If we consider the possibility of bringing valuable assets into the digital world, it makes sense that, after our identity and money, the next thing we would want to digitalize would be our title deeds. While this may seem like a distant possibility, a project I worked on with the Land Registry in 2018 demonstrated that it is not only possible but also highly desirable.

In this chapter I discuss the issues with the current system of titles, how titles could be made digitally native and the benefits of title tokenization. By the end of the chapter, you will have a solid understanding of the potential for digital titles and how they could transform the property market.

Land titles and the title register

At its simplest, a property title is a record of ownership. Having your name on a property title – or as we could also express it, owning the title – essentially means you legally own and have the right to use a piece of land. It allows you to use, sell, lease or transfer the property, although there might be specific conditions attached.

In the UK the title register plays a crucial role in property transactions. Managed by the Land Registry, this register is a detailed database that keeps track of who owns what. It breaks down into three essential parts:

- the property register, which describes the property, showing exactly where it is located and its boundaries;
- the proprietorship register, which tells you who currently owns the property and the type of ownership they have; and
- the charges register, which reveals any mortgages or legal claims against the property.

Understanding these elements is key to navigating property ownership and transactions in the UK.

The problem with titles today

The Land Registry oversees an immense portfolio of property, valued at over £8 trillion, representing a significant portion of the UK's wealth.

It is time to reconsider how we view the Land Registry. What if we thought of HM Land Registry not just as a register but as a 'digital asset custodian'? Imagine if all registered titles were assets under the custodianship of the Land Registry.

Viewed as a digital asset custodian, the service shows significant limitations. Customers are not provided with accounts and cannot directly interact with their assets, and the process of transferring these assets is marred by high costs, time-consuming procedures and a post-trade settlement process that is not only manual but also labour-intensive. The custodian of the assets can take several months to process a change of ownership (a problem known as the registration gap), and approximately 20% of digital asset transfers experience delays due to data inaccuracies (which require requisitions to obtain the correct information).

Envisioning a different future

Now, let us picture a future in which the current barriers in asset management and transfers are completely removed. Think about a world in which customers who hold assets at the Land Registry have accounts similar to bank accounts – a world in which their digital deeds are accessible and easily transferable. Imagine turning the static entries on a difficult-to-update register into a vast pool of highly liquid digital assets. This innovation would enable ownership to extend from four to an infinite number of parties, allowing for true fractional ownership. The process for transferring the assets would be quick, secure, efficient and smooth. Conveyancers would be easily able to serve their customers and interact with and support the transfer of assets, similarly to legal professionals in the securities market.

This change would do more than just cut down on costs, risks and hassles for both individuals and companies; it would also spark a surge of innovation. It would breathe new life into the property market, making it more vibrant and dynamic, and it would have a profound effect on the economy, significantly boosting GDP. This vision for the UK is not merely aspirational; it is entirely achievable. If the UK invests in and utilizes the emerging wave of tokenization infrastructure, it will have the opportunity to create a world-leading property market as part of a thriving economy and a sustainable future.

What is a digital title?

A digital title is not merely a static entry in a database, whether that entry is a scanned document or was digital data from the start. Such an entry does not constitute a digital title any more than the funds in my online bank account constitute a digital pound. For a title to be considered digital, it must possess utility, as I explored in chapter 9 on tokenization. Drawing on the financial services sector's approach to describing digital money, I propose the following definition for a digital title: 'a digitally unique, executable object that is immutable, ownable and transferable, representing ownership rights and information about a parcel of land'.

Tokenized titles with HM Land Registry

Back in 2017 I was working on a research project that explored a concept we called the 'internet of public value'. We were investigating the potential of an enterprise distributed ledger platform for the government. The research paper, published in 2018, proposed the concept of a public value network.[2] This network would use distributed ledger technology (DLT) to allow public service organizations to maintain their decentralized operations – including budgets, decision making, business and service design – while also optimizing and synchronizing these operations both locally and nationally. The aim was to encourage collaboration in designing and delivering seamless, human-centric services; automate service processes; ensure adherence to and auditing of regulations, policies and procedures; and enhance financial transparency throughout the public service value chain.

While we were deep into this research project, HM Land Registry issued a public tender looking for an R&D delivery partner to help them 'develop proof of concepts using key technologies like blockchain, distributed ledgers and smart contracts. Our goal – prove that we can make land registration and the buy-sell process easier.'[3]

Coadjute bid for the project and won. For the project, which was called Digital Street, we tokenized titles and money on a blockchain network, facilitating a seamless transfer of titles. One key innovation was that we did not seek to move the entire register to the blockchain. The central register remained the single source of truth. The title was locked on the central register, and a tokenized version of the title was issued onto the blockchain network for the purposes of the transaction. Then, after the transaction, the central register was updated, and the title on the blockchain network was 'burned'.

Figure 5 illustrates the title at the start of the process on the left and shows it moving through an on-blockchain digital ID process. In the middle the atomic swap of money and title occurs, with the title being updated to reflect the new owner. Then, at the end of the process, the central register is updated.

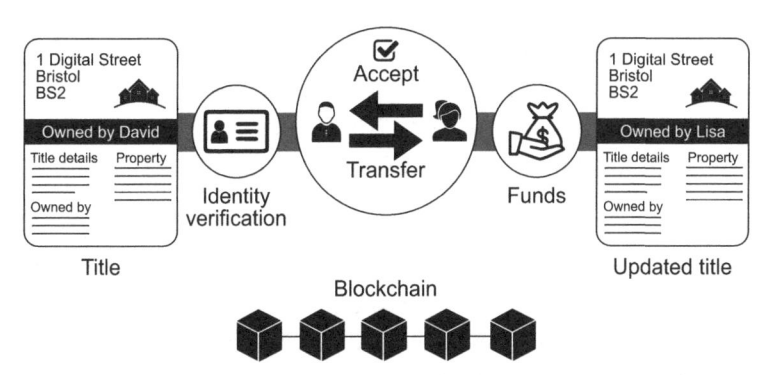

Figure 5. Transferring a title on a blockchain network. (*Source:* adapted from the Land Registry's presentation at a 2019 World Bank conference.)

The solution was showcased to the UK industry and to an international audience at the World Bank's land registration conference in Washington, receiving wonderful feedback. It was clear to me that if such an infrastructure existed it would be a significant step towards fixing the data fragmentation in the market.

Less than a year after we had successfully completed the prototype and showcased the potential for a transformative leap in property transactions through tokenization, the Covid pandemic hit the UK. This crisis dramatically shifted the priorities at the Land Registry. Instead of continuing on the path of innovation – particularly with projects that promised to redefine the property market, such as tokenizing titles – the focus had to pivot towards ensuring operational performance during an unprecedented global health emergency.

HM Land Registry leading change

Following the pandemic, HM Land Registry unveiled its strategy for 2022 and beyond, signalling a renewed commitment to the Digital Street project and a keen focus on blockchain and tokenization.[4]

This emphasis on digital innovation is highlighted early in the Land Registry's strategy document and immediately captured my interest. In the same document, Chairman Michael Mire's mention of the report co-authored by the United Nations Economic Commission for Europe (UNECE) on 'Digital transformation and land administration' piqued my curiosity, leading me to examine the report myself.[5]

The UNECE report illuminates the digital challenges that land registries face today, stressing the impracticality of clinging to outdated methods in an era marked by digital progress. It cautions that a failure to evolve could result in operational difficulties and negative feedback from the public and politicians alike.

A notable aspect of the report is the A6 model, which presents six strategies that land registries can adopt in response to digital innovations such as tokenization: 'avoid', 'analyse', 'attack', 'acquire', 'ally' and 'alternative'. The 'avoid' strategy is deemed unworkable, underscoring the risks of overlooking the movement toward digital services.

The 'analyse' strategy, while cautious, could lead to delays in embracing essential technologies, possibly leaving land administration lagging behind other industries. The 'attack' strategy, characterized by scepticism toward new technologies such as blockchain and tokenization, initially gained some traction within the land sector, despite the proven success of blockchain in finance. However, the report recommends taking a proactive stance through the 'acquire' or 'ally' strategy, suggesting the acquisition of or partnership with technological solutions to enhance land registration infrastructure:

> The evidence for the benefits of embracing digital disruption is now significant, especially post-pandemic. Land administration organizations are moving beyond this passive approach. 'Attack' involves rebutting the technological opportunity, exposing weaknesses, and downplaying the benefits of it. The initial rebuff of blockchain technologies by parts of the land sector provides a recent example, despite the technology being successfully deployed in the finance and currency sectors. This approach usually only applies short-term. 'Acquire' and 'ally' involve buying-out or buying-into a technological solution or ecosystem. Whilst some land administration organizations continue to develop in-house IT solutions, the 'acquire' and 'ally' approaches – via outsourcing and public-private partnerships – are increasingly used to scale-up IT infrastructure, services, and capacity.

If we reflect on the Land Registry's approach, it is clear that the organization is investigating how tokenization could revolutionize land registration. Given the significant opportunity presented by technology, and the advanced, operational systems now in place in financial services, there is a pressing need to take action and to progress from the analysis phase to the 'ally' phase, as advised by the UNECE report.

Current legislation supports tokenized titles

There is a strong case to be made that tokenizing titles does not require any legal or legislative overhaul. Andrew Bailey, the governor of the Bank of England, made an insightful comment regarding the tokenization of money, noting that 'the key point is that the singleness of money is preserved; it's the utility of money – what we can do with it – that changes, not the money itself.'[6] This view is applicable to property titles as well. The point is to enhance what we can do with titles through tokenization, not to alter the titles themselves.

As is made clear by the insights in the Regulated Liability Network white paper (discussed in chapter 10), adopting new technologies for managing and recording money does not require changing laws, and it is therefore reasonable to assume that the same is true of titles. The legal characterization of a title remains unchanged whether it is noted in a physical ledger, in a centralized database or on a DLT platform.

Furthermore, in 2023 the UK Jurisdiction Taskforce released a statement on the compatibility of existing English law with issuing and transferring securities on DLT systems. This statement concluded that 'the most common use cases for Digital Securities can indeed easily be accommodated within existing English law frameworks', highlighting English common law's flexibility to adapt to digital securities without needing new statutes.[7]

Property as a digital asset: lessons from financial markets

If we step outside the property market for a moment, we can see financial markets embracing digital assets, with the UK leading the way in this space. It is becoming increasingly clear that the property markets of the future will be those that utilize digital assets. To understand the feasibility and importance of this step, we can look to a recent report by UK Finance entitled 'Unlocking the power of securities tokenisation: how the UK can lead digital transformation and consolidate its role as a global financial centre'.[8]

In the foreword to the report, Bob Wigley, chairman of UK Finance, states:

> I believe the conversation around securities tokenisation is just the beginning of a much wider digital transformation – including everything from digital money to digital identification. This is a pivotal moment for the international community to build the necessary infrastructure to ensure that these innovations are rolled out safely and achieve their truest potential. This report can be a guide to shepherd us from our current moment towards a more advanced, innovative financial market, with the UK at the helm.

This view aligns with the core message of this book, which aims to contribute to moving the UK property market towards a more advanced and innovative future. There is much to learn from the developments in financial markets.

The UK Finance report begins by highlighting the growing consensus among capital markets participants that tokenization – the digital representation of real financial assets – can transform the financial system, and that the UK should be at the centre of this transformation. The report also points out that English common law already puts the UK on a strong legal footing to support digital assets, as outlined in the summaries of existing law in the UK Jurisdiction Taskforce's legal statements on cryptoassets and smart contracts, on digital dispute-resolution rules and on digital securities. The Law Commission has also shown thought leadership in its recent final report on digital assets, confirming that English and Welsh law is supportive of digital assets (including tokenized securities) within the UK.[9]

The UK Finance report sets out three missions that the UK government and industry should urgently pursue to support the development of the UK market. These missions are as follows, and each introduces activities that the UK can undertake immediately:

1. enable innovation and experimentation, underpinned by legal and regulatory certainty;
2. foster a flourishing UK digital market by promoting interoperability and safe innovation at scale; and
3. become a leader in global standards for the tokenized securities market.

A steering group focused on digital titles in the property market could adapt these missions as a way to start work on the introduction of digital titles. In relation to mission 1, the report states that HM Treasury 'should urgently roll out the first FMI [financial market infrastructure] Sandbox for the use cases identified as most pressing'. It could be argued that the Land Registry should just as urgently collaborate with HM Treasury to ensure that digital titles can be issued into this sandbox.

The report also covers in detail the benefits of tokenization, which are clearly applicable to the property market. These benefits include the following.

- *Unlocking capital.* Tokenization allows assets, including illiquid assets, to be accessed by more investors and to be traded. If sufficient liquidity can then be created, the asset will move with greater velocity through the financial system, unlocking trapped capital for investors.

- *Fractionalization.* Tokenization enables fractionalization, whereby investors can purchase fractions of an asset. This has the potential to increase access to investors (potentially including retail investors).

- *Risk management.* Because DLT can enable atomic (i.e. simultaneous and instantaneous) settlement of transactions on a 24/7 basis, tokenization has the potential to eliminate or reduce counterparty risk, bankruptcy risk and performance risk by shortening the settlement time for transactions to which two or more counterparties are bound.

One of the most interesting parts of the report is its comparison of the traditional process flow for bond purchases with the same process using tokenization. The steps and friction points of the traditional process are very closely aligned to those of a property transaction, and tokenization also has similar benefits for both. For example, the process steps in the traditional model include due diligence, placing a deposit, execution, clearing, settlement and custody. One of the benefits the report documents for the tokenized model is that 'atomic transfers can remove the need for an intermediary to clear trades, though intermediaries may still be required depending on the market' – a concept that will be explored further in chapter 14, on digital completion.

The parallels between the financial markets' adoption of digital assets and the potential for the property market to follow suit are striking. The benefits of tokenization – such as increased efficiency, transaction speed, security, liquidity, fractionalization and enhanced risk management – are just as applicable to property assets as they are to financial securities. The legal and regulatory groundwork being laid in the UK for digital assets in financial markets can serve as a template for the property market to build on.

The use of smart contracts and atomic transfers could streamline property transactions, reducing friction points and minimizing the need for intermediaries. Furthermore, the increased transparency and immutability afforded by digital assets could enhance trust and security in property transactions, reducing the risk of fraud and disputes.

The property market has much to gain from embracing digital assets. To realize these benefits, it must learn from the experiences of financial markets and adapt their strategies to property assets' unique characteristics. This will require close collaboration between industry stakeholders, regulators and technology providers to develop standards, best practices and legal frameworks that support the safe and efficient tokenization of property assets.

Conclusion

The concept of digital titles represents a significant step forward in the property market. The Land Registry aims to lead this digital transformation, as outlined in its 2022+ strategy, which renews its focus on digital innovation and tokenization. Having previously prototyped tokenized titles in 2018, the Land Registry is well positioned to take action and progress to the 'ally' phase recommended by the UNECE's report on digital transformation in land administration.

The adoption of tokenized titles does not necessitate a legal or legislative overhaul, as English common law has shown flexibility in accommodating digital securities within existing frameworks. This compatibility suggests that the property market can gain the benefits of tokenization without having to overcome significant barriers. The Land Registry can unlock new opportunities and create a more vibrant and dynamic property market for the benefit of individuals, businesses and the broader economy.

PART IV

THE DIGITAL PROPERTY MARKET

In the following chapters I explore operating in the property market, focusing on three key aspects. First, I explore how a consumer-focused 'super app' will transform the end-user experience by integrating all property-related services and information into a single, seamless platform. I then examine how businesses risk being left behind if they fail to adapt their models to the new digital ecosystem by embracing innovative, technology-driven approaches. Finally, I discuss how digital money, titles and smart contracts will revolutionize the once-cumbersome property settlement process, enabling instantaneous, secure and transparent exchanges that dramatically improve overall efficiency and reliability.

CHAPTER 12

A housing super app

> There were over 1 million residential property transactions in England in 2020/21. Despite a large number of successful transactions, the consumer experience of buying and selling property is often criticised for not being as efficient, effective, or consumer-friendly as it could be. Moving home is widely acknowledged to be among the most stressful of life experiences.
>
> — House of Commons Library research briefing, 2022[1]

In this chapter I explore the concept of a consumer 'super app'. It is reasonable to assume that consumers in the property market desire the same simplicity, certainty, transparency and security that they enjoy in other aspects of their digital lives. If we take as the starting point for this chapter the assumption that the digital foundations discussed in part II in terms of a property market infrastructure, data standards, trusted data and digital identity are in place, then it is a small next step to envisage a housing super app.

There is ample evidence to suggest that consumers want to complete the entire end-to-end property transaction on a single platform with all supporting services seamlessly integrated. This chapter sets out the rationale for a housing super app and where it might materialize in the UK property market.

Failing consumers

In the early 1990s, before the internet, the process of buying a house was a physical endeavour that required prospective buyers to visit the town of interest, personally engage with local estate agents and collect printed property details. The mortgage application process involved face-to-face meetings with bank representatives or brokers. The buyer was essentially the project manager, coordinating between various parties – estate agents, banks and conveyancers – to ensure a successful property purchase.

The advent of the internet revolutionized these processes. Websites such as Rightmove enabled buyers to browse properties online, eliminating the need for physical visits to estate agents and significantly reducing advertising and printing costs for agents. Gradually, every aspect of the property market, including HM Land Registry and payment methods, transitioned to being an electronic interface. However, the development of the property market seemed to pause after the creation of electronic interfaces and consumer portals, with no further advances to fully integrate and streamline the entire buying and selling process.

Contrast this with the evolution of the travel industry, which saw the development of aggregator sites such as Expedia and Booking.com. These platforms enabled users to plan entire holidays – from flights and accommodation to car rentals and insurance – in one seamless online experience. They combined multiple services into a single, user-friendly interface, simplifying the entire holiday booking process.

So why has the property market not experienced a similar level of aggregation? One key reason is the complexity of property transactions. Unlike booking a holiday – which involves services such as flights, hotels and car rentals that are relatively stand-alone – property transactions involve interdependent services from estate agents, conveyancers, mortgage brokers and lenders. Each service in a property transaction is intertwined with others, making it challenging to create a single platform that can effectively manage the entire process.

Aggregating a property transaction therefore requires a level of sophistication and interoperability that surpasses what is needed for less complex services. To date, the absence of the digital foundations of the market discussed earlier – such as a cohesive property market infrastructure, standardized data formats or trusted identities – has hindered the development of an integrated customer journey. If we can realize these capabilities, it will become possible to deliver an end-to-end digital customer journey akin to what has been achieved in the travel industry and other sectors.

What is a housing super app?

A housing super app is an all-encompassing digital application that transforms the home buying and selling experience by aggregating every aspect of the end-to-end process. It is designed to alleviate the burden of coordinating and integrating various services traditionally involved in property transactions. It serves as a single platform through which consumers can connect, interact and transact with all necessary parties, including estate agents, conveyancers, mortgage brokers, lenders and surveyors. Additionally, the app could extend its functionalities to encompass auxiliary services such as moving companies and insurance providers.

A good analysis of what a housing super app might look like was provided by 11:FS in their report 'Building the future of home buying'.[2] The report clearly sets out the need for businesses to collaborate in order to create a seamless customer journey: 'Businesses need to partner with adjacent businesses to serve more of their customers' needs, more effectively.'

Housing super app case studies

Before we turn to the UK market in the next section, a look at the United States and Asia will help illuminate what is happening in the super app space.

Case study: Zillow

The story of Zillow is a prime example from the United States. Founded in 2006 by Rich Barton, who drew inspiration from his experience of creating Expedia while at Microsoft, Zillow aimed to revolutionize property much as Expedia transformed travel. Interestingly, this idea came six years after the launch of Rightmove in the UK, which possibly influenced Barton's vision for Zillow.

Initially, Zillow focused on providing detailed home listings and valuations, addressing the gap Barton had noticed in online property information. Fast forward to eighteen years later, and Zillow has become a dominant force in the US property market, boasting a massive user base and substantial quarterly revenues.

In 2022 Zillow announced an aggressive strategy aimed at tripling its revenue by 2025. The company's vision involves developing a comprehensive housing application that simplifies every aspect of the property process. This includes searching for and discovering properties, renting, obtaining mortgage pre-approval, engaging in immersive property viewing, conducting in-person tours, purchasing, securing financing, selling and overseeing payments.

Zillow's February 2024 letter to shareholders clearly sets out that Zillow is going to fix the moving experience for consumers:

> Everyone who has moved knows how much time, energy and money it takes. We believe Zillow is best positioned to meet that need. To help solve this challenge, we set out to become the housing super app: a digital experience where all the disparate pieces of the gnarly moving process are integrated on one platform.
>
> The housing super app is here today. It's called Zillow.
>
> Zillow's housing super app empowers customers by delivering property data and education, a suite of Zillow-owned

solutions, and a network of best-in-class partners at their fingertips.[3]

Zillow's approach to building this super app involves various innovations. They are connecting buyers with local agents, offering integrated financing solutions and working to make property tours as simple as booking a restaurant reservation online. Their growth strategy focuses on five pillars: financing options, seller solutions, enhanced touring experiences, integrated services and an expanded partner network.

A closer look at one of these pillars, the financing process, reveals that Zillow have updated their portal to include a 'home loans' tab for pre-approval decisions and direct connections with brokers, backed by their financial services company, Zillow Home Loans, LLC. This strategy of building or acquiring supporting services was further underscored by Zillow's acquisition of the agency customer-relationship management (CRM) system Follow Up Boss in December 2023.

As Zillow ventures into new markets, it will be worth watching how its super app strategy unfolds.

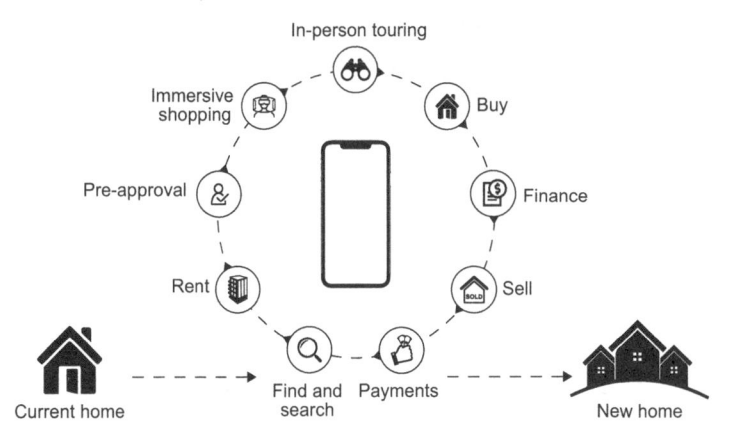

Figure 6. The Zillow housing super app ecosystem.

> **Case study: OhMyHome**
>
> OhMyHome, an innovative property portal in Asia, was established in 2016 by sisters Rhonda and Race Wong. Drawing on their personal experiences of moving more than twenty times as a family, the sisters were well versed in the challenges and aspirations of property transactions. Rhonda Wong, who became a property investor in her early twenties, encountered first hand the inefficiencies and problematic practices in the industry. This inspired her to reshape property transactions with a focus on prioritizing client interests and offering transparent services, leading to the creation of OhMyHome.
>
> Despite experiencing a decline in revenues in the first half of 2023, the company aimed to triple its revenue by 2024 and achieve positive EBITDA (earnings before interest, taxes, depreciation and amortization) by 2025. OhMyHome's strategies include evolving into a property super app.
>
> What sets OhMyHome apart from portals like Zillow is its comprehensive support for consumers who wish to sell their property independently. This positions them in competition with the agents who utilize their platform. Originally a platform serving consumers directly, it contrasts with Zillow's beginnings as an advertising portal for agents.
>
> The potential success of OhMyHome in reaching breakeven by 2025 and subsequently scaling up will be a crucial development to watch. The Asian market – as the birthplace of the world's first super app, China's WeChat – seems a natural fit for the emergence of a housing super app.

Who will build the UK housing super app?

The trend of housing super apps in both the United States and Asia raises a pivotal question about the future of the UK's property market: are we on the brink of the emergence of a housing super app in the UK, and if so, who will build it? If the United States and Asia are any indication, then the housing super app is likely to emerge from the portal space.

Zoopla as housing super app

Zoopla, established in 2007 by Alex Chesterman and Simon Kain, embarked on a mission to revolutionize the UK property market. From the outset, Zoopla aimed to distinguish itself from existing property portals by utilizing publicly accessible property data. This approach combined property searches with comprehensive research tools, empowering users to make more informed property-related decisions.

Over the years, Zoopla has emerged as the second-largest property portal in the UK. However, it holds a much smaller market share than its competitor Rightmove. Zoopla's significant milestones include its floatation on the London Stock Exchange in 2014 and its transformation into ZPG PLC in 2017. A pivotal moment came in 2018 when ZPG was acquired by Silver Lake Partners for £2.2 billion.

Zoopla has strategically expanded its portfolio by acquiring a variety of businesses. These include the CRM systems Alto and Jupix, proptech start-up YourKeys, data companies HomeTrack and Calcasa, and mortgage business Mojo Mortgages.

In 2023 Zoopla adopted a strategy that could be seen as an initial step towards a housing super app model. This involved restructuring to amalgamate its diverse acquisitions under a new brand, Houseful. The vision behind Houseful is to 'create the connections that power better property decisions', fostering a more integrated property market that enhances home choices, customer experiences and business operations. Houseful's ethos is that by uniting its brands, it unlocks the collective power of software, data and insights to advance the property industry.

This strategy focuses on acquiring and integrating various businesses to build a comprehensive digital customer experience. However, Zoopla's approach raises certain challenges. Forming a closed ecosystem, in which all high-value transactions are directed towards the primary brand's owned businesses, limits consumer choice and value. Consumers typically prefer having options, and open ecosystems that allow low switching costs usually generate more value than closed systems with restricted choices. This aspect of Zoopla's strategy could be a crucial factor in determining its long-term success and its ability to truly transform the property market experience for its users. In any event, pending legislation will give consumers the power to liberate their data from any closed ecosystem.

Rightmove as a housing super app

For over two decades, Rightmove has remained steadfastly committed to the core value proposition of being a property listing portal. However, more recently the company has started exploring new growth opportunities.

In November 2023 Rightmove announced a new vision to empower everyone with the confidence to 'make their move'.[4] This vision involves a consumer pillar, which is where the concept of a housing super app would fit. This pillar outlines the following progressive timeline.

- Present focus: becoming the go-to online platform for finding homes.

- In the next five years: evolving into a comprehensive 'moving journey assistant'.

- Beyond the next five years: aspiring to be a 'home life partner'.

This approach articulates a clear property transaction value chain, encompassing stages such as finding properties, mortgage facilitation, transaction processes (conveyancing and surveys), moving services and property life cycle management (renovations, energy and maintenance).

Interestingly, Rightmove's strategy, unlike those of Zillow in the US or Houseful in the UK, is built on an open ecosystem model. It focuses on connecting consumers with the finest agents and services available in the market, rather than acquiring businesses down the value chain to create a closed ecosystem. This approach mirrors the model of travel super apps, whereby the platform provides access to a range of services without seeking to control or acquire elements within the value chain, thus offering consumers a comprehensive choice from the entire market.

Legislative framework

Over the last five years there has been an ever-increasing legislative framework for the protection of consumer data and the creation of better and more innovative consumer digital services.

In terms of protecting consumers, the General Data Protection Regulation (GDPR), introduced in the UK in May 2018, has transformed how companies in all markets have thought about their handling of customers' data. The 2024 Data Protection and Digital Information Bill[5] will give consumers more protections, but it will also extend to other industries the type of data access piloted by the Open Banking initiative:

> The Bill enables the Secretary of State or the Treasury to make regulations requiring data holders to provide customer data either directly to a customer at their request or to a person authorised by the customer to receive the data, at the request of the customer or the authorised person.
>
> It is envisaged that data will be provided to an authorised person rather than the customer since the authorised person will be best able to make use of the data on the customer's behalf (in the provision of innovative services such as account management services via a visual dashboard of accounts, displayed on a smartphone app) but the regulation-making powers have been kept broad to allow for direct provision of data to customers in the future.[6]

Political will

There is also strong political will to enhance the consumer experience. Besides the new legislation mentioned above, this drive is evidenced by the focused research initiatives commissioned by the Ministry of Housing and the establishment of the Digital Property Market Steering Group.

Inquiry on improving the home buying and selling process

In April 2024 the Levelling Up Committee launched an inquiry on improving the home buying and selling process. The second question of the inquiry relates directly to improving the consumer experience of home buying and selling: 'How could the consumer experience be improved during the process for buying and selling homes?'[7]

The inquiry is cross-party, so it is reasonable to assume that all political parties in the UK acknowledge that the home buying and selling process is failing consumers.

Digital Property Market Steering Group

Inaugurated in the summer of 2023, the Digital Property Market Steering Group (DPMSG) is a collaborative effort to transition the market from its existing electronic form to a truly digital market. The mission of the DPMSG is to foster the adoption of digital technology in a manner that prioritizes transparency, security and consumer-friendliness.

Chaired by HM Land Registry, the DPMSG has senior representatives from a range of influential bodies, including the Law Society, the Conveyancing Association, the Council for Licensed Conveyancers, the Society of Licensed Conveyancers, the Solicitors Regulation Authority, the Royal Institution of Chartered Surveyors, the Chartered Institute of Legal Executives (CILEX),

CILEX Regulation, the Council of Property Search Organisations, Propertymark, the Building Societies Association and UK Finance. The group's vision puts improving the consumer experience at the core: 'Through collaboration, innovation and a focus on emerging digital technologies, we will build on existing progress across the home buying and selling system to get a better result for the customer: simpler, faster, more certain and less stressful.' This vision is supported by strong commitment from the organizations involved to creating a better experience for consumers. For example, Lubna Shuja, president of the Law Society of England and Wales, notes: 'The Law Society is committed to improving the home-buying experience for consumers and others involved in the conveyancing process.'

It looks like the legislative and political stages are set for the introduction of a housing super app to transform the customer journey.

The seamless property journey in the housing super app

We can therefore begin to imagine a world in which the entire home buying and selling process is consolidated into a single, intuitive digital platform – a true housing super app for property. This is the reality that homeowners and prospective buyers will soon be able to enjoy, revolutionizing the way they navigate one of life's biggest financial and logistical challenges.

For sellers, the journey could begin by simply opening the app and accessing the comprehensive 'how to sell' guide. Here, they find clear, step-by-step instructions on preparing their property for the market, including reviewing and updating their digital property logbook. This repository of verified information about the home is continuously updated in real time and is instantly accessible to the agent they select within the app.

With a few taps, the seller can onboard their chosen agent, who then guides them through the next stages. The app provides a transparent breakdown of all anticipated moving costs, while also facilitating secure digital identity checks. Crucially, these identity credentials are securely shared across the ecosystem, eliminating redundant paperwork.

For leasehold properties, the app's integration with management companies ensures swift and affordable access to the necessary documentation. This level of coordination streamlines the entire process and saves sellers valuable time and frustration.

Buyers enjoy an equally seamless experience within the super app. They can review government guidance on home purchasing, access listings and even initiate the mortgage process – all without leaving the platform. The app's intelligent matching algorithm connects buyers with suitable lenders based on the property's material information, removing the guesswork.

Once an offer is accepted, the app seamlessly transfers the complete set of property details to the buyer's conveyancer, kick-starting efficient due diligence. Throughout this stage, buyers can track progress, communicate with their legal team and arrange surveys from within the housing super app environment.

Finally, the app facilitates a frictionless completion process. Funds are transferred securely and prioritized through the banking system, ensuring a timely, stress-free transaction. Utility accounts and property access are automatically updated, minimizing disruption for all parties.

The housing super app will transform the home buying and selling journey. By consolidating all the necessary services and information into a single, integrated platform, homeowners and prospective buyers will enjoy the same levels of convenience, transparency and control that they do in other aspects of their digital lives. This digital-first approach will transform what is today a complex and frustrating process into a seamless, stress-free experience.

Conclusion

The current property market is ripe for the development of a housing super app that coordinates and simplifies the home buying and selling experience. The fragmented nature of the property transaction process, with each business vying for customer attention on their respective platforms, has resulted in a suboptimal customer experience.

Case studies from the United States and Asia, such as Zillow and OhMyHome, demonstrate the potential of housing super apps to revolutionize the property market by offering seamless, end-to-end services. In the UK both Zoopla and Rightmove are positioning themselves to potentially become the UK's housing super app, with varying strategies and approaches.

The legislative framework and political will to enhance consumer experiences and data protection further support the development of a housing super app in the UK. The establishment of the Digital Property Market Steering Group and its commitment to improving the consumer experience signify a strong push towards a more digital, streamlined and consumer-centric property market.

As the property market continues to evolve, the emergence of a housing super app in the UK appears increasingly likely. Such an app will transform the way consumers navigate the property transaction process and set new standards for convenience, efficiency and simplicity in the industry.

CHAPTER 13

The business ecosystem

> In the future, competition will no longer be about products, it will be about ecosystems.
>
> — Geoffrey A. Moore

For businesses, the current online property market revolves primarily around their own portals or platforms, offering them a sense of security, control and ownership. However, as the market evolves, these spaces may soon become secondary to the main property transaction hubs. This chapter suggests that the property market will follow a similar digital maturity curve as other sectors, such as payments, with an increasing number of services becoming 'embedded' within other platforms.

While many businesses, particularly those upstream of portals and estate agents, may feel threatened by the embedded service model, this concern is often unfounded. The shift towards embedded services presents an opportunity for growth and innovation, as demonstrated by the success of embedded finance models.

In this chapter I explore the concept of an embedded service strategy, which seeks to create seamless, fully integrated digital journeys for customers. By transitioning from a stand-alone model to an embedded service model, businesses can unlock new value for both consumers and themselves, surpassing the limitations of the stand-alone or intermediary panel model.

177

At the end of this chapter, you should have a clear understanding of the embedded service model and how to develop an effective embedded service strategy for your business.

Transactions are getting progressively slower

Consumers and businesses increasingly expect property transactions to smoothly integrate a full suite of services. From a consumer's standpoint, the entire process of buying or selling property should ideally take place on a single platform, offering easy access to services provided by conveyancers, brokers, surveyors, lenders and others, all working together seamlessly.

Yet an examination of the property market over the past two decades uncovers an unexpected trend. Despite technological advances that have expedited transactions across nearly every other sector, the property market has seen a significant extension in transaction times, as indicated by HM Land Registry data. This phenomenon prompts a critical question: why is the property market witnessing a decline in efficiency in an age dominated by digital innovation and technological advances?

Friction from intermediaries

As discussed in chapter 3, the internet spawned the rise of panel businesses. Initially online marketing businesses, these would work with, say, estate agents to capture leads and sell them to downstream businesses on their 'panel', such as brokers or conveyancers, for a referral fee. In my experience, the growth of panel businesses that operate as pure marketing intermediaries between estate agents and conveyancers within the market has had a profound impact on the property transaction process. Contrary to their initial purpose as facilitators for streamlining property transactions, these businesses have actually brought about a decrease in both the speed and efficiency of transactions, directly affecting the profitability of conveyancers.

Furthermore, these panel businesses often levy substantial fees on businesses they refer or pass leads to. For example, conveyancers can pay fees of up to £400 per case – with a portion of this going rightfully to the estate agent, as effectively a sales commission, but often with an outsized portion going to panel intermediaries, which impacts profitability and quality of service. These outsized returns are often at odds with the value they generate as intermediaries, and the true value creators, such as agents and conveyancers, are squeezed on fees, with quality of service coming under pressure. For this very reason, the referral fees charged by panel businesses have been under significant scrutiny from the government and Trading Standards over the years, with the 2024 Levelling Up Committee inquiry into home buying and selling making panel fees a key area of investigation.[1]

From my vantage point, there is a clear causation, not just a correlation, between the rise of panel businesses and the decline in transactional efficiency. Rather than simplifying the process, these businesses have introduced more complexity and bureaucracy, leading to a slowdown and fragmentation that affects everyone involved. At the heart of this issue lies the irony that panel businesses, set up to improve efficiency, have instead compounded inefficiencies.

This reduction in efficiency in the property market is indicative of a larger problem: the failure of panel businesses to innovate effectively. True innovation should focus on streamlining processes, removing barriers and enhancing value for all stakeholders. Consequently, the property market stands at a critical juncture, necessitating a redesign to create efficiency and better serve its customers. This shift is essential for the market to regain its lost momentum and continue providing valuable services to all parties involved.

Embedded services outperform panel services

The property market is currently undergoing a transformation that focuses on what customers truly want: a seamless integration of all necessary services in a property transaction. The outdated practice of directing customers between multiple service providers is becoming obsolete. Modern consumers are gravitating towards a unified experience, with every required service being directly integrated into their transaction journey, eliminating the need to manage multiple log-ins and disjointed processes.

Many professionals in the property sector – such as conveyancers, brokers and others who traditionally operate upstream of estate agents and portals – show resistance to this integrated model. However, today's customers are less willing to navigate through various separate portals. They prefer starting their journey in one place, with all the necessary services smoothly incorporated into their chosen digital environment. This preference highlights a growing gap between traditional service delivery methods and the streamlined, customer-centric approach now in demand.

For primary and secondary service providers in the property market, adopting embedded services offers a strong economic rationale. By doing away with the inefficiencies and costs associated with fragmented customer journeys brokered by panel companies, and instead creating an efficient, streamlined experience based on embedded services, these providers can significantly cut down on referral fees.

Additionally, they can engage directly with other parties involved in the transaction, fostering a fairer distribution of value. For instance, if a conveyancer wants to avoid hefty panel fees, it can integrate its services into an estate agent's platform, resulting in a win–win situation.

The model of directly embedded services not only simplifies the transaction process but also elevates the value delivered to customers. It aligns with customer convenience and efficiency, strengthening the competitive stance of adopters. In a market increasingly driven by customer experience and convenience, those who successfully integrate their offerings into a cohesive, unified customer journey are poised to benefit in terms of both customer satisfaction and financial viability.

The property market is at a critical juncture, with directly embedded services set to revolutionize service delivery standards. By adopting this integrated approach, conveyancers, brokers and other service providers can better align with customer expectations, reduce operational costs and create a more engaging, efficient and value-driven customer journey. This shift represents not just a technological advance but a fundamental change in conducting property transactions, emphasizing convenience, efficiency and a focus on the customer.

Embedded service strategies

Brokers, conveyancers, lenders, surveyors and other market participants are now finding themselves navigating the accelerating transition towards a digital property market. To succeed in this environment, they must not only deeply understand market dynamics but also identify and capitalize on their unique strengths to tap into new distribution avenues.

The journey towards a successful embedded service strategy begins with a critical evaluation of one's role within the market. Market participants must decide whether they aim to be the primary consumer platform, akin to portals and estate agents, or to position themselves as downstream players, seamlessly integrating their services into offerings provided by upstream providers. This decision will have a profound impact on their overall strategy and the nature of their interactions within the market.

Next, market participants must identify the starting point of the customer journey. This requires a deep understanding of customer behaviour and preferences, as well as a keen eye for identifying the optimal point to integrate their services. By pinpointing where customers begin their journey, businesses can ensure a seamless experience that meets customer needs and expectations.

Equally important is the evaluation of one's value proposition. This involves a critical analysis of the value offered to both customers and upstream distribution partners. To thrive in the digital property market, businesses must shift away from traditional thinking that focuses on being the primary customer touchpoint. Instead, they should consider how they can enhance the customer experience through embedded services that complement and augment the offerings of their partners.

In pursuit of this goal, valuable insights can be gleaned from studying digital journeys and embedded services in other sectors, such as travel and embedded payments. These industries have been at the forefront of merging various services into a unified customer journey, offering valuable lessons that can be adapted and applied to the property market.

Building brand affinity and customer loyalty is another crucial aspect of an effective embedded service strategy. Even if a business is not the primary brand at the distribution point, there are still ample opportunities to foster brand loyalty. This can be achieved by consistently delivering exceptional service throughout the customer life cycle and by improving transaction flow for upstream partners through enhanced service speed and quality.

However, the success of an embedding strategy hinges on a business's digital maturity. It is essential to conduct a thorough self-assessment of one's digital readiness, ensuring that the necessary software and application programming interfaces (APIs) are in place to integrate services seamlessly into partners' platforms.[2]

Embracing a new business model is also necessary to adapt to the digital market. The transition from a traditional online property market to one that aligns with a digital-first, AI-enhanced approach requires a fundamental rethinking of established business models.

By carefully evaluating these various aspects, market participants can develop an embedding strategy that not only meets current market demands but also remains adaptable to future changes. The key lies in focusing on service integration and creating cohesive customer journeys that improve the overall value proposition and strengthen one's digital market presence. This marks a significant shift from competition based on individual services to a more collaborative, integrated model that prioritizes customer experience and operational efficiency. In the digital property market, success will increasingly depend on the ability to seamlessly integrate services and provide a frictionless experience for customers.

To achieve this, market participants must be willing to collaborate and share data in ways that may have been uncommon in the past. This requires a shift in mindset from a siloed, competitive approach to one that is more open and cooperative. By working together to create a unified customer journey, businesses can unlock new opportunities for growth and innovation. Of course, this transition is not without its challenges. Integrating services and sharing data raise concerns about security, privacy and data ownership. It is essential for market participants to address these issues proactively and develop robust frameworks for data governance and protection. Moreover, the shift to an embedded service model may require significant investments in technology and infrastructure. Businesses must be prepared to invest in the necessary APIs, software and platforms in order to enable seamless integration and data sharing.[3]

Despite these challenges, the benefits of an embedded service strategy are clear. By providing a more integrated, efficient and customer-centric experience, businesses can differentiate themselves in an increasingly competitive market and build lasting relationships with customers. As the digital property market continues to evolve, embracing an embedded service strategy will be essential if brokers, conveyancers, lenders, surveyors and other market participants are to remain competitive and relevant. By understanding market dynamics, evaluating their capabilities and capitalizing on their unique strengths, these businesses can position themselves for success in the digital, AI era.

The journey towards an embedded service model may not be easy, but it is a necessary one. As customer expectations continue to rise and technology advances at an unprecedented pace, those who can adapt and innovate will be best positioned to thrive in the years ahead. Ultimately, the success of an embedded service strategy will depend on the willingness of market participants to collaborate, share data and create a more open and integrated ecosystem. It will require a fundamental shift in mindset and a commitment to putting the customer at the centre of everything they do.

But for those who are willing to change, the rewards are significant. By providing a more seamless, efficient and personalized experience, businesses can build stronger relationships with customers, differentiate themselves from competitors and unlock new opportunities for growth and innovation. As the digital property market continues to evolve, one thing is clear: an embedded service strategy is no longer a nice-to-have, but a necessity for survival. Those who can successfully navigate this transition will be the ones who shape the future of the industry and reap the benefits of a more connected, efficient and customer-centric market.

Avoiding service customization

In discussions with various conveyancers, I have found that a recurring apprehension is the potential commoditization and fee reduction from integrating their services within upstream channels. This concern, however, misses a crucial point about the current market situation, where intermediaries are often taking significant fees while also driving up operational costs and are less efficient than connecting directly to partners. Understanding the economic logic behind embedded services requires recognizing their multiple advantages. Direct collaboration with entities such as estate agents allows conveyancers to go beyond their conventional transactional function. This collaboration leads to a smooth, digitally integrated customer journey, thus improving the overall service value. The crux of this approach is in simplifying and streamlining processes, and thereby removing the complex layers that currently encumber property transactions.

A key benefit of this model is the reduction of intermediary fees. This removes a drain on financial resources and allows for a reallocation of funds. This reallocation is more than just cost-saving; it is a strategic investment in better service quality. Funds saved from intermediary fees can be reinvested into technology, enhancing the customer experience and improving processes. This investment leads to a more efficient and customer-focused service model, thereby increasing the perceived value of the services provided.

This new paradigm can counteract the current trend of declining returns. In a scenario in which conveyancing services are seamlessly integrated into the customer journey, the value delivered to consumers significantly increases. This higher value enables conveyancers to justify their fees based on the actual value they add to the transaction.

Embedded services also create an environment in which customer satisfaction and loyalty are crucial. Happy customers are more likely to return and recommend services, generating a positive cycle of business growth. In this environment, conveyancers can use their superior service delivery as a competitive edge, attracting more customers and securing a stable revenue stream.

Conclusion

The property market is on the cusp of a significant transformation, driven by the need to improve efficiency and better serve customers. The current market comprises fragmented services and suffers from lengthy transaction times, and it is ripe for disruption through embedded service models. This new approach, which has proved successful in other sectors, offers a seamless and integrated customer journey by eliminating the complexities and bureaucracy associated with multiple intermediaries.

By adopting an embedded service strategy, businesses in the property market can unlock new value, reduce operational costs and create more engaging, efficient and value-driven customer experiences. This shift represents a fundamental change in conducting property transactions, emphasizing convenience, efficiency and a focus on the customer.

To successfully navigate this transition, businesses must assess their digital maturity, understand their role in the new ecosystem and build brand affinity through exceptional service. By learning from other markets and embracing a collaborative, integrated model, the property market can regain its lost momentum and deliver the streamlined, digital-first experience that modern consumers demand.

CHAPTER 14

Digital completion

> Project Meridian was a highly collaborative effort between the Bank of England, HM Land Registry, the project's vendor – Coadjute, and the Bank for International Settlements (BIS) Innovation Hub London Centre.
>
> — Project Meridian report, 2023[1]

In this chapter I explore how digital completion, powered by tokenized money and titles along with smart contracts, can transform property transactions. Today's manual process for completing property transactions is laden with inefficiencies, including delays, a high risk of errors and fraud and a general lack of transparency. By harnessing digital innovation and the emerging infrastructure of regulated financial services, we can start on the path to overcoming these challenges and reshaping how property transactions are executed.

The pioneering Project Meridian – a collaborative project between the Bank of England, the Bank for International Settlements, HM Land Registry and Coadjute – developed a prototype synchronization service connected to the Bank of England's real-time gross settlement (RTGS) system and the Land Registry. The project provided a service whereby conveyancers could request the simultaneous transfer of money and a land title. The project was a success and proved the case for synchronization and automation within property transactions, also known as digital completion.

The process and benefits defined in the Project Meridian report can now be delivered through tokenized money and titles, as well as the deployment of smart contracts for programmable transactions. These capabilities will soon be provided by the UK Regulated Liability Network and the Mastercard Multi-Token Network.

This chapter starts by addressing the challenges in the current market and ends with a blueprint for the digital completion of property transactions. It outlines the benefits for banks, conveyancers, estate agents, the Land Registry and, most importantly, consumers.

Transaction completion today

Let us look at an example transaction to better understand the complexity. A typical property transaction might require £500,000 to settle. These funds may come from various sources: £350,000 from a mortgage, £100,000 from savings and £50,000 from the buyer's parents. Then, this £500,000 will need to be distributed among several parties: for example, £200,000 to the outgoing mortgage lender, £5,000 in stamp duty to HMRC, £5,000 in estate agent fees and the remaining £290,000 to the seller. A transaction in which multiple fund sources and destinations are involved will be slow and complex, since it needs to be manually orchestrated as a series of individual payments.

A significant part of the responsibility for successfully moving all these funds around manually falls to the buyer's conveyancer, who plays a crucial role in facilitating the exchange of funds and ensuring a smooth settlement process. One of their primary responsibilities is to collect all the necessary funds into their client account, holding them in escrow until the completion date.

The task of coordinating multiple sources of funds can be particularly challenging when there are multiple buyers involved, such as a husband and wife. In that case, the manual process requires close communication with the buyers and their mortgage lender, as well as meticulous record-keeping to track the various contributions.

Settling the transaction

Prior to collecting all the completion funds, the buyers conveyancer will do an 'official search of whole with priority' via the Land Registry – this effectively locks the title for thirty days.[2] Once the buyer's conveyancer has successfully collected all the necessary funds, they proceed to coordinate with the seller's conveyancer in order to schedule a date and time for the completion of the transaction. The completion process (think settlement) then involves the buyer's conveyancer transferring the funds to the seller's conveyancer.

As soon as the seller's conveyancer confirms receipt of the funds, they release the deeds to the buyer's conveyancer, officially completing the transaction. However, this crucial step is still performed manually, leaving room for potential risks such as manual errors, delays or business emails being compromised.

Dispersing funds

After receiving the funds from the buyer's conveyancer, the seller's conveyancer is responsible for dispersing the money to the appropriate parties and accounts. Prior to the completion day, the seller's conveyancer contacts the seller's mortgage lender to obtain a repayment figure for the outstanding mortgage. They must also be prepared to adjust the repayment figure if the completion date moves out, meaning they are repaying the mortgage at a later date.

Conveyancers often report that obtaining a redemption figure from mortgage lenders can be a cumbersome and time-consuming process. In addition to dealing with the lender, the conveyancer must also obtain and carefully verify the seller's bank account details to ensure that the remaining funds are sent to the correct account. This step is critical to ensuring the accuracy of the fund disbursement. As well as redeeming or obtaining discharges for all mortgages, the responsibilities of the seller's conveyancer at this stage involve sending the transfer documents and any title deeds to the buyer's conveyancer.

Post-completion

The responsibilities of the buyer's conveyancer continue even after the completion of the transaction. They are tasked with lodging the necessary stamp duty land tax (SDLT) forms and paying any SDLT due. Once they receive the transfer documents, they must lodge the application for registration at the Land Registry within the thirty-day period of the official search.

To minimize the likelihood of the Land Registry raising requisitions,[3] the buyer's conveyancer carefully reviews the application, ensuring that all required documents are properly dated, executed and attached. They also address any discrepancies in names between the deeds and the register, either by resolving them or providing a clear explanation.

After submitting the application, the buyer's conveyancer sends a copy of the title information document to the buyer and reminds them to keep their service address updated. They also inform the client about the Land Registry's Property Alert service, highlighting its benefits in protecting against property fraud.

Once the registration process is complete, the buyer's conveyancer notifies both the buyer and any lender involved. Additionally, they handle other important documents, such as mortgage loan agreements, planning permissions and indemnity policies, following the lender's instructions meticulously. This ensures that all aspects of the transaction are concluded accurately and efficiently.

The completion, settlement and post-completion stages of a property transaction require meticulous attention to detail, effective communication and a high level of coordination from the conveyancers. By carefully managing the flow of funds, verifying account details and handling the necessary paperwork, conveyancers play a vital role in ensuring the smooth transfer of property ownership.

Problems with today's manual completion process

Picture this: you are in the process of buying your dream home, and after months of searching, negotiating and finalizing the details, you are finally ready to complete the transaction. However, instead of a smooth and straightforward process, you find yourself having to transfer life-changing amounts of money to the account of a conveyancer that you have never transferred money to before, and you then have to wait and hope that all the other money arrives with the solicitor and transfers successfully to the seller.

This manual completion process, which sees a small army of 4,000 conveyancers moving more than £300 billion each year, has to be one of the most inefficient settlement systems in the world.

One of the most significant drawbacks of the manual completion process is the high risk of errors and fraud. With so many parties involved, including buyers, sellers, conveyancers and mortgage lenders, the potential for miscommunication and mistakes is alarmingly high. A simple typo in a bank account number or a misplaced decimal point can derail the entire transaction, causing delays and financial losses for all parties involved.

Moreover, the reliance on email communication and physical paperwork makes the process vulnerable to sophisticated email scams and fraud. Imagine receiving an email from what appears to be your conveyancer, requesting that you transfer your deposit to a new account. You follow the instructions, only to later discover that the email was a fake and your hard-earned money is gone. Unfortunately, such scenarios are not uncommon, and with sophisticated AI they will become more prevalent.

Another major issue with the manual completion process is the lack of transparency and visibility. As a buyer, seller or business involved in the transaction, you are often in the dark, unsure of the progress of the transaction or at what time it will be completed. You may spend hours on the phone trying to get updates from your conveyancer, or trying to reschedule removal companies.

If a consumer is selling their current home to buy a new one, they may find themselves in a precarious position, uncertain of when they will receive the funds from the sale and whether they will arrive in time for them to complete their purchase. This uncertainty can have a ripple effect, impacting their ability to move in on the planned day.

The inefficiencies of the manual completion process also extend to the allocation of funds. During the transaction, large sums of money are often held in escrow accounts, where they sit idly, not earning interest or being put to productive use.

Bank of England and HM Land Registry collaboration

It seems to me a significant gap in our national infrastructure that there is no direct digital link between our sovereign money system and our sovereign land system.

In 2023 fate gave me the opportunity to bridge this gap by suggesting a collaboration between the Bank of England, the Bank for International Settlements and the Land Registry. The idea was to use the housing market as a test case for a proposed synchronization service. The proposal was successful, leading to Project Meridian.

Project Meridian

In August 2018 the Bank of England issued a 'call for interest' regarding the delivery of proposed synchronized settlement functionality as part of its RTGS Renewal Programme. The core capability of a synchronization service is 'atomic settlement':

this involves linking the transfer of two or more assets in such a way that one asset is transferred if, and only if, the other asset (or assets) is also transferred.

Responding to strong demand for such functionality, the Bank of England, in partnership with the Bank for International Settlements, embarked on developing a prototype for this synchronization service in 2023. The project – named Project Meridian – aimed to explore and refine the business case for synchronization in practical scenarios.

Coadjute was selected to develop this prototype, which focused on the housing market to demonstrate the potential for more efficient and automated transactions. I led Coadjute's efforts.

The Meridian synchronized settlement solution demonstrated a fully digital process for completing transactions, aligning with a digital adaptation of the Law Society's completion protocol. It begins with counterparties appointing a synchronization operator, who initiates the transaction and ensures the buyer's funds are reserved specifically for this purpose. A crucial step involves placing an earmark on these funds, preventing their use elsewhere for a designated period until approval is obtained.

Once all conditions are met, the synchronization operator orchestrates the transfer of funds from the buyer's bank to the seller's through the RTGS system. This pivotal moment also triggers the updating of both the buyer's and seller's bank account balances to reflect the transaction's completion. Finally, participants receive confirmation of the settlement, and the digital deed is lodged with HM Land Registry, facilitating the update to the register.

Digital completion using tokenized commercial bank money

Building on Project Meridian's foundation, I have further explored digital completion in property transactions through both the Mastercard MultiToken Network platform and the Regulated Liability Network. Both of these projects (described in chapter 10) involved replicating Project Meridian's synchronized digital completion process for the movement of funds via tokenized commercial bank money.

From here it is not a big step to imagine expanding the Regulated Liability Network to accommodate not just digital money but digital titles as well. This extension would enable atomic settlement (the instant, simultaneous exchange of title deeds and money), thereby enhancing transactional smoothness and reducing the risk of defaults or delays.

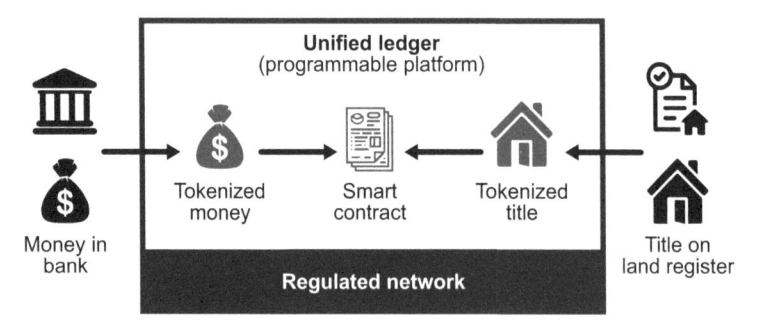

Figure 7. Digital completion using tokenized money and titles.

Figure 7 outlines the following steps of the digital completion process.

1. Commercial banks issue tokenized money.

2. The Land Registry issues tokenized titles.

3. Titles are committed to a smart contract.

4. Money is committed to a smart contract.

5. The transaction executes. The smart contract is tailored to execute the token exchange based on specific conditions,

such as the completion date. The transaction is atomic, ensuring delivery-versus-payment certainty – that is, either all money and titles transfer or nothing transfers.

6. Funds and titles are then off-ramped, with bank ledgers and the Land Registry updated in real time.

The benefits of digital completion

The integration of tokenized money and titles within an escrow-style smart contract significantly improves the efficiency, security and transparency of the property transaction process, bringing a multitude of benefits for all stakeholders involved, including consumers, conveyancers, lenders and estate agents. With this technology, the property industry can revolutionize the way transactions are conducted, making the process more efficient, secure and convenient for everyone.

For consumers, one of the greatest advantages of automation is the flexibility it introduces. Currently, there is a campaign in the UK property market to ensure completions happen by 13:00 on moving day. With digital completion, not only can this objective be programmed, but the process can be taken a step further by enabling a 24/7, 365-day completion window, as opposed to today's 9:00–14:00, Monday to Friday limitation. The ability to move houses even over a weekend would provide a level of convenience previously unimaginable.

Moreover, the approach to reserving funds in digital completion allows consumers to continue earning interest on their money right up until the moment of completion. This feature ensures that buyers and sellers maximize their financial returns throughout the process, as their funds are not sitting idly in escrow accounts for extended periods.

In addition to the financial benefits, automation also brings a heightened level of security to property transactions by effectively eliminating the risk of push-payment fraud. This is particularly important given the high value of property transactions and the devastating consequences that fraud can have on individuals and families.

For conveyancers, the ability to reserve and conditionally release funds through an automated system grants them an unprecedented level of security and control over transactions, minimizing the effort and risks associated with manual cash handling. By streamlining their operations and reducing the potential for human error, conveyancers can focus on providing their clients with high-quality legal advice and support, rather than getting bogged down in administrative tasks. Furthermore, the lowered risk profile of automated transactions is likely to be reflected in reduced professional indemnity insurance costs for conveyancers.

Lenders also stand to benefit greatly from the adoption of automation in property transactions. By retaining mortgage funds and customer deposits up until the moment of completion, lenders can improve their capital efficiency and reduce their exposure to risk. The introduction of 'straight-through' processing and automated workflows further enhances these benefits, reducing costs and minimizing the potential for errors or delays.

Moreover, the increased payment transparency and traceability that comes with automation provides lenders with a clearer picture of their financial position and helps them to make more informed decisions about their lending practices. This can lead to better risk management, improved customer service and, ultimately, a more stable and sustainable lending environment.

Estate agents, as key facilitators of property transactions, also have much to gain from the move towards automation. Like other stakeholders, they benefit from the increased payment transparency and traceability that the system provides. This not only makes their job easier but also helps to build trust and confidence with their clients. Perhaps most importantly, the instant payment on completion that automation enables is a major advantage for estate agents in an industry where cash flow is king.

Beyond the financial benefits, the streamlined and transparent approach that automation brings to property transactions also has a positive impact on the overall customer experience. By making the process smoother, faster and more secure, estate agents can provide a higher level of service to their clients, building stronger relationships and increasing the likelihood of repeat business and referrals.

The benefits for HM Land Registry

The transition to a digital property market faces notable hurdles within the current land registration system, including data silos, registration delays, frequent requisitions and cumbersome interactions. However, the advent of tokenization, which allows for the digital representation of property titles and funds on a blockchain, has emerged as a groundbreaking solution, particularly when combined with the evolving infrastructure of regulated financial services.

Tokenization paves the way for a more integrated and efficient system by enabling property titles and funds to exist in a digital format on a blockchain. This development not only facilitates a seamless connection with financial institutions and transaction orchestration services but also eradicates the issue of data silos. With a single, reliable source of truth accessible in real time, all parties involved can swiftly verify information. The implementation of smart contracts automates transactions, allowing data to flow effortlessly between interconnected systems. This transition to straight-through processing eradicates manual verification, significantly enhancing the speed, accuracy and reliability of property transactions.

One of the most pressing challenges in the current system is the registration gap – the delay between the execution of a property transaction and its official registration. Tokenization directly eliminates this gap by enabling the immediate, atomic update of title tokens at the moment of the transaction. This ensures that the legal and beneficial titles are updated concurrently with the transaction, minimizing the risk of discrepancies and subsequent legal complications.

Furthermore, tokenization significantly reduces the frequency of requisitions from the Land Registry. By incorporating Land Registry data at the outset of the transaction process through an orchestration service, potential discrepancies are identified and rectified in real time, rather than post-transaction. This pre-emptive approach diminishes the administrative and financial burden associated with rectifying errors.

Another critical issue within the current framework is the inefficiency of interactions caused by the absence of an account-based system or user identity service at the Land Registry. Tokenization introduces a solution through digital wallets for property owners, functioning as accounts for their tokenized titles. These digital wallets enable secure and straightforward property transactions, akin to those in online banking, and eliminate the need for intermediary involvement. Property owners gain the ability to directly manage transactions or authorize third-party services to act on their behalf, markedly reducing the current system's friction and inefficiency.

Conclusion

The integration of tokenized money and titles, combined with smart contracts, has the potential to revolutionize property transactions. The current manual completion process is fraught with delays, high risk of errors and fraud, increased costs and stress, inefficiencies, dependencies and a lack of transparency. By embracing digital innovation and using the emerging regulated financial services infrastructure, we can address these challenges and transform the way property transactions are conducted.

The collaboration between the Bank of England, the Bank for International Settlements and the Land Registry through Project Meridian demonstrates the potential for synchronization and automation in property transactions. By adopting tokenized money and titles, and by utilizing smart contracts for programmable transactions, we can streamline processes, enhance security and provide a more efficient and transparent experience for all parties involved. This shift towards truly digital completions of property transactions will ultimately benefit consumers, conveyancers, lenders and the broader economy.

CHAPTER 15

Smart contracts

> Smart legal contracts could revolutionise the way we do business, particularly by increasing efficiency and transparency in transactions. We have concluded that the current legal framework is clearly able to facilitate and support the use of smart legal contracts; an important step in ensuring increased recognition and facilitation of these agreements. Our related work on digital assets and conflict of laws will further establish England and Wales as a global leader for technological innovations in the digital sphere.
>
> — Professor Sarah Green, Law Commissioner, 2021[1]

In this chapter I explore the opportunity for the digital transformation of legal contracts, with a particular focus on property transactions. Over the course of my thirty-year career, contracts have remained largely unchanged, consisting mainly of Word documents. The property market involves not only the primary contract of sale but also numerous ancillary contracts for supporting services, presenting a significant opportunity to streamline and enhance the home buying and selling process. By digitalizing contracts, we can align them with the other digital components of property transactions, such as identity, money and titles.

I will examine the challenges associated with contracts today, as well as the advantages of smart legal contracts and their legal standing. By the end of the chapter you will have a comprehensive understanding of smart legal contracts, their features and the potential they hold to transform property transactions.

The history of contracts

Contracts have been integral to human society for millennia, facilitating trade and cooperation by formalizing agreements between parties. The history of contracts can be traced back to ancient civilizations. These early contracts were primarily used for trade, marriage and employment agreements, serving as binding promises that were often witnessed by officials and sealed with a clay seal or signature.

In ancient Egypt, contracts were used in complex construction projects and to delineate property rights, while in ancient Greece, contracts began to take on more diverse forms, including agreements for services, loans and partnerships. The Romans further developed contract law, introducing concepts such as *consensus ad idem* (agreement to the same thing), which remains a foundation of contract law today.

Medieval Europe saw the expansion of contract usage beyond local communities, facilitated by the growth of trade fairs and the merchant class. The Lex Mercatoria, or merchant law, emerged as a body of commercial law that governed trade across Europe, relying on common practices and principles recognized by merchants.

The Industrial Revolution and the growth of international trade in the nineteenth century necessitated more complex contracts and the development of modern contract law. Legal systems around the world began to standardize and codify contract law, establishing the principles that govern contractual agreements today, such as offer and acceptance, consideration, capacity and legality.

In the twentieth century, technological advances and globalization further evolved contract law. Digital contracts and electronic signatures have become commonplace, enabling faster and more efficient transactions across borders.

What is a contract?

In English law a contract is essentially an agreement between two or more parties that is legally enforceable. This means that if you make a deal with someone to exchange something of value – like goods, services or money – and you both agree to certain terms, then you have a contract. For a contract to be valid under English law, it must contain several key ingredients.

The first ingredient is offer and acceptance. One party makes an offer, and the other accepts it. This can be as simple as agreeing to buy a book for a certain price.

The second ingredient is consideration. Both parties must exchange something of value. For example, one person pays money, and the other delivers a product or service.

The third ingredient is the intention to create legal relations. Both parties must intend that their agreement will be legally binding. This means they expect the law to support their agreement if anything goes wrong.

The fourth and final ingredient is capacity. The parties involved must have the legal ability to enter into a contract, meaning they are of sound mind, not under duress and of a certain age (usually over 18).

Contracts can be made in writing or verbally or even be implied by actions, but some types of contracts must be written down to be valid, such as those involving the sale of property. If someone does not do what they agreed to in the contract, the other party can ask a court to enforce the agreement or seek compensation for any losses caused.

The problem with contracts today

Contracts are a vital part of property transactions, from the initial listing of a property to the final transfer of ownership. They provide clarity, build trust and give legal protection to all parties involved. Estate agents' contracts, conveyancing service agreements, mortgage broker contracts, mortgage agreements and property sale contracts are just a few of the many contracts that form the backbone of the property market. As the market grows and evolves, the volume and complexity of these contracts continue to increase.

However, despite their importance, the way we approach contracts in the property market has remained largely unchanged for decades. Most property professionals still rely on decades-old software such as PDF and Microsoft Word to create and store agreements in a digitized format, but these are essentially just digital copies of paper documents. This approach fails to take advantage of modern technology that allows for dynamic data access and streamlined contracting.

One of the most significant problems with traditional contracts in the property market is the inefficiency of the manual processes involved. Consider a typical property sale: the estate agent captures the property information, the seller's details and the buyer's details. The seller's conveyancer will then have to recapture all this information when creating the contract pack for the buyer's conveyancer, who, in turn, will manually review it. This duplication of effort is a clear example of the inefficiencies that plague the property transaction process.

The manual nature of these tasks, from data entry to contract review, is time-consuming and prone to errors. For instance, a conveyancer may make a typo when re-entering the property address or the buyer's name, which can lead to confusion and delays down the line. Similarly, manually reviewing lengthy contract packs that are in PDF or Word format can be tedious, and it is easy for important details to be overlooked, such as a specific clause or a deadline.

One positive aspect is that for most sales the Law Society's Standard Conditions of Sale contract is used, so there is an opportunity for a standardized data model to sit under this (see chapter 7, on trusted data).

Limited accessibility and collaboration are also significant hurdles in property transactions. In any industry where paper-based or word-processed contracts are sent over email, it is challenging for multiple parties to access, review and collaborate on the same document simultaneously. This is particularly an area of inefficiency in property transactions, as they involve multiple stakeholders, including two conveyancers, a buyer and a seller. Emailing back and forth different versions of a contract can be confusing and time-consuming and can lead to version-control issues.

Security and fraud prevention are also critical concerns in property transactions. Traditional paper-based contracts can be easily lost, damaged or tampered with, while contracts in the form of a Word document can be easily altered or intercepted when exchanged by email over the internet. This is particularly worrying given the high value of property transactions and the sensitive personal and financial information involved. Fraudulent activities such as forged signatures, altered contract terms and even people selling properties that are not theirs can have devastating consequences for buyers and sellers alike.

Finally, tracking and enforcing contractual obligations in property transactions is manual with traditional contracts. Manually keeping track of deadlines, payments and milestones across multiple contracts is time-consuming and error-prone for all parties. It is all too easy for a crucial deadline to be missed, a payment to be delayed or a contractual obligation to be overlooked. This can lead to delays in the transaction process, financial losses and even legal disputes.

What are smart legal contracts?

The Law Commission defines smart legal contracts as

> a legally binding contract in which some or all of the contractual obligations are defined in and/or performed automatically by a computer program. Smart contracts, including smart legal contracts, tend to follow a conditional logic

with specific and objective inputs: if 'X' occurs, then execute step 'Y'. Smart legal contracts are expected to increase efficiency and certainty in business, and to reduce the need for the contracting parties to have to trust each other; the trust resides instead in the code.[2]

Building on this definition, we can consider the two main features of smart legal contracts to be

- that some or all of the contractual obligations under the contract are performed automatically by a computer program ('automaticity'), and
- that the contract is legally enforceable.

HM Land Registry smart contracts

HM Land Registry first identified the potential of self-executing contracts for conveyancing in its Chain Matrix project in 2007. This envisaged an automatic exchange of contracts when every condition for the completion of each transaction in the property chain had been met.[3]

In March 2019 the Land Registry conducted a proof-of-concept digital property transfer using blockchain and smart legal contracts. The transfer conducted to test the prototype was a simulation of a previously concluded property transaction that had taken 22 weeks to complete.

The test digital transfer was concluded in under 10 minutes – dramatically below the benchmark. As each stage of the transfer was completed using blockchain technology, the parties could access an automatically updated record of the transaction, where they could see all actions that had been taken and what actions still needed to take place to complete the transaction. The successful transfer demonstrated that smart legal contracts and blockchain technology can enable property transactions that are speedier, more secure, more transparent and trusted by the parties involved.[4]

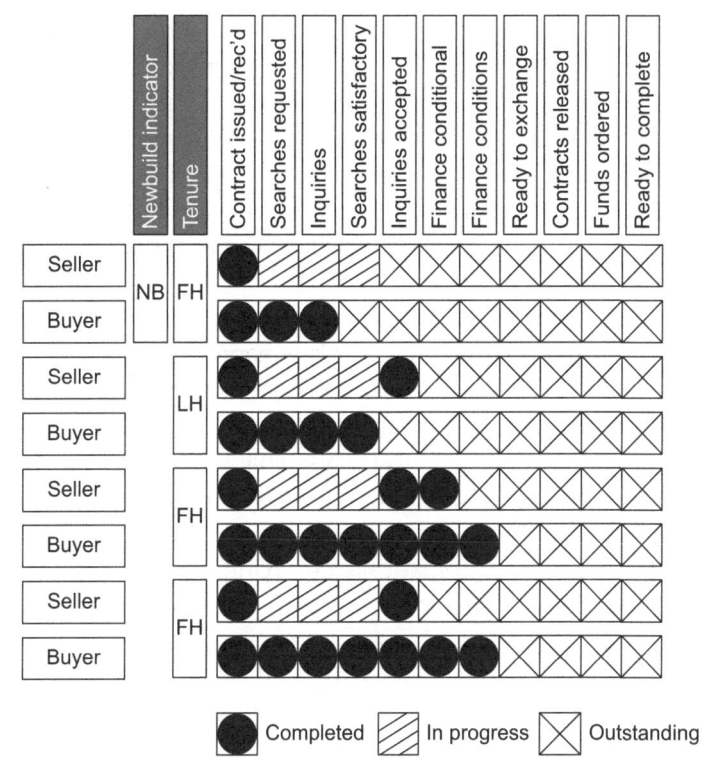

Figure 8. An illustration of the Land Registry's Chain Matrix system, showing the transactions in the chain (the rows) and the actions required for their completion (the columns). Also shown are indicators of whether the property is a newbuild (NB) and whether it is leasehold or freehold (LH or FH).

The benefits of smart contracts

The benefits of smart contracts in the property sector are numerous and have the potential to revolutionize the way transactions are conducted. In 2021 I worked with LawtechUK and its UK Jurisdiction Taskforce to create a report on smart legal contracts in the home buying and selling process.[5] Looking at use cases in the UK and internationally, the report concluded that there were numerous benefits to be gained from the adoption of smart legal contracts.

One of the key benefits is the automation and efficiency that smart legal contracts bring to property transactions. By automating tasks such as transferring funds and registering titles, smart contracts streamline the process, reducing the need for manual intervention and minimizing the risk of human errors. This automation not only speeds up transactions but also results in lower transaction costs and an improved user experience for all parties involved.

Smart contracts also enhance security and help prevent fraud in property transactions. This enhanced security reduces the risk of tampering and unauthorized access, protecting the interests of buyers, sellers and other stakeholders involved in the transaction. Smart contracts also enable the real-time monitoring of contractual obligations – such as payment schedules, inspections and document submissions – throughout the transaction process. By automatically enforcing contract provisions when predefined conditions are met, smart contracts ensure the timely completion of tasks, streamline dispute resolution and minimize the need for costly and time-consuming legal interventions. This automation and real-time tracking provide greater transparency and accountability in the transaction process, reducing the potential for delays and disputes.

The benefits of smart contracts in the property sector are not just theoretical. The LawtechUK report highlighted several use cases for which smart contracts have been successfully implemented, both in the UK and internationally. These use cases demonstrate the practical applications of smart contracts and the tangible benefits they can bring to the property sector.

The Law Society's Standard Conditions as smart legal contracts

Another significant advantage of smart contracts is the standardization and interoperability they offer in property transactions. Smart contracts can be standardized for specific types of transactions, such as residential or commercial property sales. This standardization simplifies the process of creating and executing contracts, ensuring consistency across transactions and facilitating seamless integration with other digital services, such as property databases, financial systems and regulatory platforms. The ability to integrate with other systems is crucial in creating a more efficient and connected property ecosystem.

The Law Society's Standard Conditions of Sale serve as a foundation for residential property transactions in England and Wales, providing information standards and clear guidelines through various operational clauses. As mentioned above, transforming this contract into a smart legal contract could have significant benefits. Let us take the example of automating one of the operational clauses. The 'deposit requirement' clause mandates the buyer to pay a deposit as a commitment to the transaction. In the traditional approach, managing this clause involves manual processes, such as collecting, holding and transferring the deposit.

Using smart contracts, the deposit requirement clause can be automated, streamlining and securing the process. Upon signing the digital contract, the smart contract can automatically reserve the deposit funds in the buyer's bank account. Once the required amount is reserved, the smart contract notifies both parties and holds the funds until completion. Upon completion, the smart contract automatically releases the deposit to the seller. In the case of a cancellation, the deposit is released back to the buyer.

This automation of the deposit requirement clause demonstrates the potential for smart contracts to simplify and secure various aspects of the property transaction process. By reducing manual interventions and automating the transfer of funds based on predefined conditions, smart contracts can significantly reduce the risk of errors, delays and disputes. Moreover, the automated process provides transparency and certainty for both parties, as they can easily track the status of the deposit and rely on the smart contract to execute the transfer according to the agreed-on terms.

Other operational clauses – such as those on formation of the contract, description of the property, ownership and planning permissions, setting the completion date, breach and dispute resolution, notices and communication, and risk, insurance and liability – could also benefit from smart contract automation. In doing so, they would make residential property transactions more efficient, cost-effective and secure, while reducing the need for manual intervention.

Legal enforceability

The Law Commission and LawtechUK have confirmed that smart legal contracts, which are executed entirely or partially in code, are legal under English law. They have found that, with some adaptations, current laws can apply to smart contracts just as they do to traditional contracts. This means that England and Wales have a legal framework flexible enough to support smart contract innovations without needing new laws. This makes the region an ideal place for businesses looking to use smart legal contracts, which promise to enhance efficiency and innovation in various sectors. The Law Commission states:

> We have concluded that the current legal framework is clearly able to facilitate and support the use of smart legal contracts. Current legal principles can apply to smart legal contracts in much the same way as they do to traditional contracts, albeit with an incremental and principled development of the common law in specific contexts. In general, difficulties associated with applying the existing law to smart legal contracts are not unique to them, and could equally arise in the context of traditional contracts. Our findings therefore build on the conclusions of the Legal Statement issued by the UK Jurisdiction Taskforce ('UKJT'), which

establishes that the current legal framework is sufficiently robust and adaptable so as to facilitate and support the use of smart legal contracts. The flexibility of our common law means that the jurisdiction of England and Wales provides an ideal platform for business and innovation, without the need for statutory law reform.[6]

Conclusion

The digital transformation of contracts, particularly in property transactions, is an essential step towards modernizing and streamlining the process. The history of contracts shows a clear progression from ancient civilizations to the present day, with contracts evolving to meet the needs of society. However, in recent years, traditional contracts have remained largely unchanged, relying on outdated software and manual processes.

The introduction of smart legal contracts addresses the challenges associated with traditional contracts, such as inefficiency, a lack of standardization, limited accessibility, security concerns and difficulty in tracking and enforcing. By automating and digitalizing contractual obligations, smart legal contracts offer numerous benefits, including increased efficiency, standardization, enhanced security and real-time tracking.

The successful trial of smart legal contracts by the Land Registry demonstrates the potential for speedier, more secure and more transparent property transactions. Furthermore, the legal enforceability of smart contracts under English law provides a favourable environment for the adoption and innovation of smart contracts in various sectors. As we move towards an increasingly digital future, embracing smart legal contracts in property transactions is a crucial step in transforming the industry and improving the overall experience for all parties involved.

PART V

THE INTELLIGENT MARKET AND THE PROPERTYVERSE

Today, all the conversation in the UK property market is about its digital future. The industry is abuzz with discussions around the digitalization of processes, the emergence of new digital assets and the formation of a comprehensive digital ecosystem. However, just as these digital foundations are being laid, an even more profound transition is appearing on the horizon.

CHAPTER 16

Artificial intelligence

> AI will have a more profound impact on humanity than fire, electricity and the internet.
>
> — Sundar Pichai, CEO of Alphabet

Artificial intelligence has come a long way: from its early, narrow applications such as playing chess, to a transformative technology reshaping industries at an unprecedented pace. I am convinced that this disruption will significantly impact the property and mortgage markets, and faster than most people anticipate.

The introduction of computers, the internet and online transactions was largely a gradual process led by individual firms. Compared with financial services and payments, the property market still has significant ground to cover in establishing basic digital foundations. However, AI is not something that can be postponed or ignored. Its impact on every person, business and stakeholder within the market will be swift and profound. When faced with deep-fake customer representations, escalating cyber-attack sophistication and competitors achieving substantial improvements in operational efficiency, a wait-and-see approach is no longer viable.

In this chapter I look at AI's history and its current capabilities. This will set the scene for the next chapter, in which I will explore a future intelligent market.

What is AI?

According to John McCarthy, widely considered the godfather of AI, artificial intelligence is

> the science and engineering of making intelligent machines, especially intelligent computer programs. It is related to the similar task of using computers to understand human intelligence, but AI does not have to confine itself to methods that are biologically observable. Intelligence is the computational part of the ability to achieve goals in the world. Varying kinds and degrees of intelligence occur in people, many animals and some machines.[1]

In brief, then, AI encompasses a wide range of techniques that enable machines to perform tasks intelligently.

The evolution of AI

In the summer of 1956 a small group of scientists met at Dartmouth College. Their meeting marked the beginning of AI as we know it today. They believed that machines could be made to mimic human intelligence, and this idea would be the foundation for decades of innovation in the field.

The late twentieth and early twenty-first centuries marked a transformative period in AI, setting the stage for a series of dramatic advances that not only showcased AI's capabilities but also symbolized its potential to surpass human expertise in specific domains.

One of the most iconic moments in this era was the defeat of world chess champion Gary Kasparov by IBM's Deep Blue in 1997. This event was a cultural turning point that shifted the public's perception of AI. Deep Blue's victory demonstrated the power of AI in processing and evaluating millions of positions, using brute computational strength to outmanoeuvre one of the greatest minds in chess. This achievement underscored the potential for AI to perform at and beyond human levels in structured, rule-based environments.[2]

AI's ability to match and surpass human expertise continued with the victory of DeepMind's AlphaGo over Go champion Lee Sedol in 2016. Go, an ancient board game known for its deep strategic complexity, had long been considered a bastion of human cognitive superiority. AlphaGo's win was significant not just for the complexity of the task but for the way the AI system approached problem-solving. Using deep neural networks and reinforcement learning, AlphaGo was able to mimic human intuition, making moves that baffled human experts but ultimately secured victory.

This landmark event highlighted the advances in machine learning and AI's ability to learn, adapt and innovate beyond the capacity of its human creators.[3]

The current era of AI, starting in about 2017, is characterized by remarkable achievements in natural language processing (NLP) and other domains, defining the cutting edge of AI research and applications. The publication of the paper 'Attention is all you need' by Ashish Vaswani and his team in 2017 marked a pivotal moment in the field of AI, particularly in the way machines understand and process human language.[4] This groundbreaking research introduced the 'transformer' model, which revolutionized the approach to NLP, set new benchmarks for machine comprehension of language and catalysed a rapid explosion of AI innovations in the years that followed.

Understanding the scale and pace of these advances is crucial for property market professionals to formulate an appropriate response to AI. The succession of rapid advances since the introduction of the transformer model have driven significant changes across industries, and the property market is no exception. These innovations underscore the urgent need for property market professionals to adapt to and embrace AI technologies in order to stay competitive and efficient in their fields.

Generative AI capabilities

As AI continues to evolve at an unprecedented pace, it is crucial for the property market to proactively engage with these technologies and understand their capabilities. There is no doubt that generative AI is a groundbreaking technology that will revolutionize all industries, including the property market and its various sectors, whether estate agency, conveyancing, mortgage lending or surveying.

Below are some of the core foundational capabilities of generative AI. I set out these capabilities in generic terms, and in the next chapter I will explore the specific impacts they might have on the property market.

In customer service, generative AI can be leveraged to create intelligent chatbots and virtual assistants. These AI-powered systems can engage in natural, human-like conversations with customers, providing instant support, answering queries and resolving issues. By analysing customer inquiries and generating contextually relevant responses, generative AI can significantly enhance the customer experience, reduce response times and alleviate the workload of human customer service representatives.

Marketing and advertising can also benefit greatly from generative AI. By analysing customer data, preferences and behaviour, generative AI can create personalized marketing content, such as product descriptions, social media posts and email campaigns. This targeted approach can lead to higher engagement rates, increased conversions and improved customer loyalty. Additionally, generative AI can assist in creating visually appealing and compelling advertisements by generating images, videos and animations that resonate with the target audience.

In the field of product design and development, generative AI can streamline the ideation and prototyping processes. By learning from existing product designs and user preferences, generative AI can generate novel and innovative product concepts, accelerating the brainstorming phase. Moreover, generative AI can create realistic 3D models and simulations that enable designers to visualize and test their ideas virtually, thus reducing the need for physical prototypes and saving time and resources.

Generative AI can also transform the way businesses approach content creation and management. From generating engaging blog posts and articles to creating product descriptions and user manuals, generative AI can automate the content creation process. This not only ensures consistency and quality across various platforms but also frees up valuable time for content creators and marketers, who can then focus on strategic tasks and creative direction.

In software development, generative AI can revolutionize the coding process. By learning from vast repositories of code and understanding programming patterns, generative AI can automatically generate code snippets, suggest optimizations and even complete entire functions. This can significantly accelerate development cycles, reduce coding errors and improve overall code quality. Generative AI can also assist in tasks such as code documentation, testing and debugging, enhancing the efficiency and productivity of software development teams.

Generative AI can also play a crucial role in predictive maintenance and quality control. By analysing sensor data and historical maintenance records, and thereby predicting when equipment is likely to fail, generative AI can enable proactive maintenance and reduce downtime. Additionally, generative AI can detect anomalies and defects in manufacturing processes, ensuring higher product quality and minimizing waste.

In the financial industry, generative AI can be applied to fraud detection and risk assessment. By learning from vast amounts of financial data and identifying patterns, generative AI can detect suspicious activities, such as fraudulent transactions or money laundering attempts. This can help financial institutions proactively mitigate risks, protect their customers and comply with regulatory requirements.

Finally, generative AI can revolutionize the way businesses approach simulation and testing. By generating realistic and diverse scenarios, generative AI can enable comprehensive testing of products, systems and algorithms. This is particularly valuable in the automotive, aerospace and manufacturing industries and others in which physical testing can be costly and time-consuming. Generative AI-powered simulations can accelerate development cycles, reduce costs and improve the overall quality and safety of products.

Conclusion

The rapid advances in AI, particularly generative AI, are poised to have a profound and transformative impact on the property and mortgage markets. The evolution of AI – from its early beginnings in the 1950s to current state-of-the-art models such as GPT-4, DALL-E and PaLM – has showcased the technology's immense potential to revolutionize industries and reshape the way we live and work.

As the property and mortgage markets continue to navigate the challenges of the digital age, it is imperative for professionals in these fields to recognize the significance of AI and proactively engage with its capabilities. By harnessing the power of generative AI, businesses can streamline processes, enhance customer experiences, drive innovation and unlock new opportunities for growth and success.

CHAPTER 17

The intelligent market

> Recent McKinsey Global Institute forecasts suggest that generative AI could offer a boost as large as $17.1 to $25.6 trillion to the global economy, on top of the earlier estimates of economic growth from increased work automation. They reckon that the total impact of AI and other automation technologies could produce up to a 1.5–3.4 percentage point rise in average annual GDP growth in advanced economies over the coming decade.
>
> McKinsey & Company, 2023[1]

Just as the internet revolutionized the way we communicate, access information and conduct business, AI is poised to have an equally profound impact. As I explored in chapter 3, the arrival of the internet transformed the property market by making information more accessible, enabling online listings, dematerializing money, titles and identities, and enabling online communication between buyers, sellers and property professionals. Now, as we stand on the brink of the AI revolution, the property market is once again facing a technological shift that will redefine how properties are bought, sold and managed.

There are significant parallels between the internet's impact on the property market and the potential impact of AI. Just as the internet made property information more widely available, AI is set to provide even deeper insights and personalized recommendations to buyers and sellers. Where the internet enabled virtual tours, AI can take this a step further by generating immersive, interactive experiences that allow potential buyers to explore properties in unprecedented detail. And just as the internet streamlined communication, AI-powered platforms will enable seamless, intelligent interactions between all stakeholders in the property transaction process.

However, the transformative potential of AI goes beyond simply enhancing existing processes. AI has the power to fundamentally alter the way decisions are made, risks are assessed and value is created in the property market. By using vast amounts of data and advanced machine-learning algorithms, AI can uncover hidden patterns, predict market trends and optimize complex decision-making processes in previously impossible ways.

Of course, realizing the full potential of AI in the property market will require more than just technological innovation. It will require close collaboration between stakeholders – from property professionals and technology providers to policymakers and regulators – to ensure that the benefits of AI are maximized while the risks are mitigated. It will require a willingness to rethink traditional ways of doing things and to invest in the skills and infrastructure needed to support an AI-driven future.

But the opportunities are immense. As we stand on the brink of the AI revolution, the property market is poised for a transformation that will create new sources of value, unlock hidden efficiencies and ultimately reshape the way we buy, sell and manage one of our most important assets. Just as the internet changed the game in the late twentieth century, AI is set to define the property market of the twenty-first century and beyond.

By providing a holistic view of AI's potential influence on the UK property market, this chapter aims to help professionals, policymakers and property market stakeholders think about how to navigate the AI-driven future of the industry.

Every sector of the market will be impacted

AI will have a transformative impact on every sector within the property market, from the way properties are searched and valued to how transactions are conducted and managed. By automating time-consuming tasks, providing data-driven insights and enhancing decision making, AI can streamline processes, reduce costs and improve the overall efficiency of the property market.

In this section I will take the generic AI capabilities from the last chapter and apply them to the various professions within the market, in order to imagine where value might be created and how the AI future might look.

AI in estate agency

AI-powered algorithms will continue to transform the way estate agencies handle property searches and provide recommendations to their clients. By using machine-learning techniques, property portals and estate agents can analyse vast amounts of data – including property features, location, price and historical sales records – to identify patterns and insights that might not be immediately apparent to human agents. This will enable estate agencies to provide more accurate and personalized property recommendations, tailored to their clients' individual preferences and requirements. For example, an AI-powered property search tool could learn from a client's browsing history, saved searches and feedback to refine its understanding of their ideal property, presenting them with increasingly relevant options over time. By continuously learning and adapting, these AI algorithms can significantly enhance the property search experience, saving clients time and effort while increasing their likelihood of finding the perfect home.

One of the most significant ways in which AI will impact estate agencies may be the deployment of AI-powered chatbots and virtual assistants. These intelligent systems will provide round-the-clock support to clients, answering common queries, offering guidance and even facilitating property viewings and transactions. By using natural language processing (NLP) and machine learning, chatbots will engage in human-like conversations, understanding the context and nuance of client inquiries and providing relevant, personalized responses. This not only enhances the customer experience by providing instant, 24/7 assistance but also frees up human agents to focus on more complex and high-value tasks, such as going out and winning listings.

AI in conveyancing

One of the most significant applications of AI in conveyancing will be the automation of document review and contract generation using NLP. Conveyancing involves dealing with a plethora of legal documents, such as contracts, deeds and certificates, and reviewing and drafting these manually can be time-consuming and prone to human error. By using NLP techniques, AI systems will analyse and interpret the content of these documents to extract relevant information and identify potential issues or discrepancies. This automation will not only speed up the document review process but also ensure a higher level of accuracy and consistency. Moreover, AI-powered tools for document generation and form filling could create standardized, error-free documentation such as HM Land Registry submissions, further streamlining the conveyancing process and reducing the risk of legal complications.

Due diligence is a critical aspect of conveyancing, involving the thorough investigation of a property's legal, financial and physical status. Traditionally, this process has been labour-intensive and heavily reliant on human expertise. However, AI will transform the way due diligence is conducted by enabling the analysis of vast amounts of data and the identification of patterns that may not be immediately apparent to human professionals. By using machine-learning algorithms and data-mining techniques, AI systems can quickly and accurately assess a wide range of information sources, such as property titles, planning permissions, environmental reports and financial records. This

AI-powered analysis can uncover potential risks, liabilities or opportunities associated with a property, which conveyancers can then use to make more informed decisions and provide better advice to their clients. Additionally, AI can help identify fraud or money laundering attempts by recognizing suspicious patterns in transaction data, further enhancing the security and integrity of the conveyancing process.

AI in mortgage broking and lending

AI algorithms will revolutionize the way mortgage brokers and lenders provide advice and recommend products to their clients. These algorithms will take into account a wide range of factors, such as a borrower's financial history, employment status, credit score, risk profile, affordability and preferences – along with data on the property, such as risk assessments. By analysing these vast amounts of data, AI algorithms can generate highly personalized mortgage recommendations tailored to each individual's unique circumstances. This not only saves time for both the borrower and the broker but also ensures that the recommended products are well aligned with the borrower's needs and financial goals.

The mortgage application process has traditionally been a time-consuming and paper-intensive endeavour, requiring borrowers to submit extensive documentation and lenders to manually review and verify the information provided. However, AI will transform this process through powerful data extraction and analysis capabilities. By using technologies such as optical character recognition (OCR) and NLP, AI systems will automatically extract relevant information from digital documents, such as bank statements, payslips and tax returns. This not only reduces the need for manual data entry but also minimizes the risk of errors and inconsistencies. Moreover, AI algorithms will analyse the extracted data in real time, identifying patterns, anomalies and potential red flags that may require further investigation. This streamlined approach to data extraction and analysis will enable mortgage lenders to process applications more efficiently, reducing turnaround times and improving the overall customer experience.

Mortgage fraud is a significant concern for lenders, as it can result in substantial financial losses and reputational damage. Detecting and preventing fraud has traditionally been a challenging task, requiring manual review of loan applications and supporting documentation to identify inconsistencies, misrepresentations and other red flags. AI will provide mortgage lenders with increasingly powerful tools to combat fraud. When these tools not only use the data provided directly to the lender but also draw on the databases of others via a shared market infrastructure, they can quickly analyse vast amounts of information – including loan applications, credit reports and property records – to identify potential instances of fraud. These systems can detect subtle patterns and anomalies that may be indicative of fraudulent activity, such as inconsistencies in income or employment information, suspicious property valuations, or unusual transaction patterns. By flagging these potential issues in real time, AI-powered fraud detection systems enable lenders and all professionals in the property transaction to investigate and address concerns promptly, reducing the risk of losses and protecting the integrity of the lending process.

AI in surveying

AI will continue to transform the way property inspections and data collection are conducted in the surveying industry. By using AI-powered tools and sensors, surveyors can automate many aspects of the inspection process, reducing the time and effort required while improving accuracy and consistency. For example, drones equipped with high-resolution cameras and AI-powered image-recognition algorithms can quickly and efficiently capture detailed aerial imagery of properties, identifying features such as roof condition, landscaping and potential hazards. Similarly, AI-enabled sensors can be deployed within buildings to monitor various aspects of the property, such as structural integrity, energy efficiency and indoor air quality. These sensors can continuously collect data, providing surveyors with real-time insights into the property's condition and performance. By automating data collection and analysis, AI enables surveyors to focus on higher-value tasks, such as interpreting the data and providing expert advice to clients.

AI-powered data analysis is transforming the way surveyors approach risk assessment and due diligence in the property market. By leveraging advanced analytical techniques, such as pattern recognition and anomaly detection, AI systems can quickly identify potential risks and issues that may impact a property's value or suitability for a particular use. For example, AI algorithms can analyse large datasets – such as environmental reports, planning documents and legal records – to uncover any restrictions, liabilities or compliance issues associated with a property. Similarly, AI can assess the risk of natural disasters, such as floods or earthquakes, by analysing historical data and predictive models. By providing a more comprehensive and data-driven understanding of a property's risk profile, AI enables surveyors to make more informed decisions and provide better advice to their clients. This not only helps to mitigate potential losses but also streamlines the due diligence process, saving time and resources.

AI at the level of the market: transforming transactions

Once AI has been deployed within organizations to optimize them internally, it will be deployed more and more at the level of the market. We can expect estate agents, brokers, conveyancers, mortgage lenders and the Land Registry to deploy AI agents to support and streamline the transaction process – all connected via a shared market infrastructure. These AI agents would work together to improve the efficiency and performance of the multi-organizational process of home buying and selling.

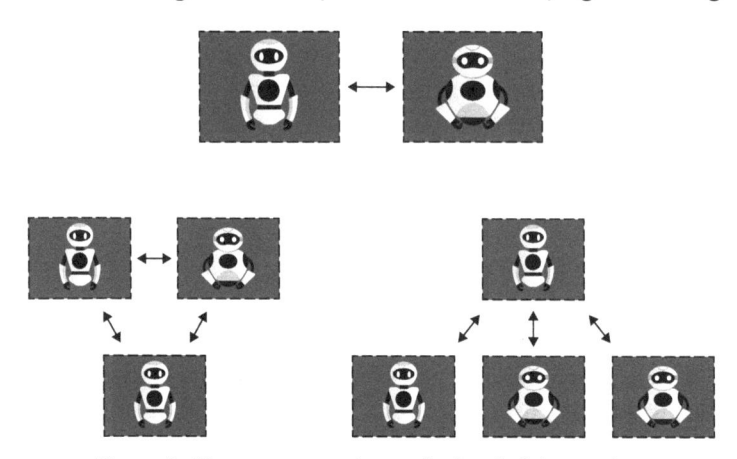

Figure 9. AI agents operating at the level of the market.

Imagine a scenario in which a potential buyer expresses interest in a property. The estate agent's AI agent would immediately engage with the buyer, answering their initial questions, providing detailed property information and even scheduling a virtual tour of the property. The AI agent would then assess the buyer's preferences, financial situation and other relevant factors to determine their suitability for the property and likelihood of completing the purchase.

Simultaneously, the mortgage lender's AI agent would be working in the background, analysing the buyer's financial data, credit history and other relevant information to determine their eligibility for a mortgage. The AI agent would then generate personalized mortgage offers based on the buyer's unique circumstances, presenting them with the most suitable options.

Once the buyer decides to proceed with the purchase, the conveyancer's AI agent would step in to handle the legal aspects of the transaction. The AI agent would automatically generate and review the necessary contracts and documents, ensuring they are accurate and compliant with all relevant laws and regulations. The AI agent would also conduct thorough due diligence on the property by analysing data from various sources to identify any potential issues or risks.

Throughout the process, all of the AI agents would be in constant communication with each other, sharing relevant information and updates in real time. This would ensure that all parties are kept informed of progress and that any potential issues are identified and addressed promptly. For example, if the conveyancer's AI agent discovers a potential issue with the property's title, it would immediately notify the AI agents of the estate agent and mortgage lender, allowing them to take appropriate action.

As the transaction progresses, the AI agents would continue to optimize and streamline the process, automating tasks where possible and providing intelligent recommendations to the human stakeholders. For example, the mortgage lender's AI agent might suggest alternative financing options if the buyer's circumstances change, while the conveyancer's AI agent could recommend ways to expedite the legal process.

Finally, when all the necessary checks and processes are complete, the AI agents would coordinate with the Land Registry to ensure a smooth and efficient transfer of ownership. The Land Registry's AI agent would verify all the relevant documentation and data, ensuring that everything is in order before registering the transfer of ownership.

With AI agents deployed at the level of the market, rather than just within individual firms, the entire property transaction process would become more efficient, transparent and customer-centric. The AI agents would work together to optimize the end-to-end process, reducing delays, minimizing errors and providing a seamless experience for buyers and sellers.

The potential impact on jobs

The accelerating pace of AI innovation highlights the urgency for both public and private sectors within the property market to prepare for the impending changes. To stay competitive, embracing these new technologies is essential.

A comprehensive study conducted by researchers from Princeton, the University of Pennsylvania and New York University has shed light on how AI, and particularly language models such as ChatGPT, may influence various occupations and industries.[2] The study introduces the AI occupational exposure methodology (AIOE), a scoring system designed to quantify this influence.

The AIOE score evaluates the relevance of AI capabilities to fifty-two work-related skills, drawing on a wide survey as well as sources such as the Electronic Frontier Foundation and the Occupational Information Network, managed by the US Department of Labor. This score reflects both the prevalence and importance of these skills within a job, thereby gauging how AI might affect it.

The study's findings reveal that many core professions within the property market are among the top 100 jobs most exposed to AI. These include telemarketers, loan officers, lawyers, insurance sales agents, personal financial advisors, property brokers and paralegals. The term 'exposure' here refers to how AI might transform these jobs, either by substituting human tasks or augmenting human capabilities.

A key takeaway from the 2023 World Economic Forum's Growth Summit was: 'AI won't take your job, but somebody using AI will.' This statement underscores the importance for property market professionals – particularly those in brokering, conveyancing and lending – to become proficient in AI tools. Below are some recommendations.

- Firstly, it is crucial for property market professionals to prioritize learning about AI tools and applications that are relevant to their field. Gaining an understanding of how AI can be utilized in their specific area of work will not only improve their work efficiency but also elevate their value within the job market.

- In addition to technical know-how, professionals should focus on developing skills that complement AI rather than those at risk of being replaced by it. Skills such as critical thinking, creativity and interpersonal abilities are facets of professional expertise that AI cannot easily replicate. By enhancing these skills, individuals can ensure they add unique value that augments AI capabilities rather than competes with them.

- The emergence of AI in the property market is likely to create new roles and demand for expertise. Professionals should be open to adapting to these new opportunities, which might include roles such as AI system managers, data analysts focusing on AI-generated insights or consultants who use AI to provide enhanced customer services. Adapting to these roles may require additional training and a flexible mindset, but it can also open up new career paths within the evolving market.

- In roles such as estate agency, for which the importance of personal relationships and trust cannot be overstated, the significance of human-centric skills remains undiminished. Professionals in these areas should strive to enhance their ability to build strong relationships, deeply understand client needs and deliver personalized services. These skills ensure a competitive edge that AI cannot overshadow and will continue to be highly valued by clients seeking a personal touch.

- Finally, staying informed about the latest developments in AI within the property market is essential. This continuous learning will enable professionals to anticipate market changes, adapt their strategies proactively and use AI to their advantage.

A bright AI future for buyers and sellers

When it comes to the property market, the transaction and completion process is often the most stressful and time-consuming aspect for buyers and sellers. However, the emergence of an

AI-powered property market holds the promise of transforming this experience, making it faster, simpler and more certain for all parties involved.

One of the key ways AI will revolutionize the transaction process is by automating and streamlining various stages, from the initial property search to final completion. AI-powered platforms will guide buyers and sellers through the process, collecting necessary information, verifying identities and performing essential checks, such as anti-money-laundering and know-your-customer procedures. By automating these steps, AI will significantly reduce the time and effort required, allowing transactions to progress more quickly and efficiently.

AI will play a crucial role in gathering and verifying property information upfront, such as title deeds, planning permissions and energy performance certificates. Using advanced techniques such as NLP and computer vision to extract and analyse relevant data from documents, AI can minimize the need for manual input and reduce the risk of errors and omissions. This upfront information gathering will provide buyers with a clearer picture of the property they are considering, enabling them to make informed decisions more quickly and with greater confidence.

AI will also streamline the communication and coordination between the various parties involved in a property transaction, such as buyers, sellers, estate agents, conveyancers and mortgage lenders. By automating the sharing of information, updates and milestones, AI will ensure that everyone stays informed and aligned throughout the process. This improved communication will help to minimize delays, reduce the risk of misunderstandings and ultimately lead to a faster and more efficient transaction.

Another significant benefit of AI in the transaction process is its ability to enhance the accuracy and speed of property valuations. AI algorithms can analyse vast amounts of data – including historical sales, local market trends and property features – to provide highly accurate and objective valuations in a matter of minutes. This not only helps buyers to make more informed offers but also ensures that sellers receive a fair price for their property. By reducing the time and uncertainty associated with property valuations, AI will contribute to a more streamlined and efficient transaction process.

Finally, AI has the potential to revolutionize the completion process, making it faster, simpler and more certain for buyers and sellers. Smart contracts, powered by AI and blockchain technology, can automate the transfer of funds and property ownership once all the necessary conditions have been met. This automation will reduce the reliance on manual processes and paperwork, minimizing the risk of errors and delays. By providing a secure, transparent and efficient means of completing transactions, AI will bring greater speed, simplicity and certainty to the final stages of the property buying and selling process.

Conclusion

AI's ability to process vast amounts of data at unprecedented speeds allows it to automate time-consuming tasks, provide deeper insights and offer personalized recommendations – all of which fundamentally enhance the decision-making process for buyers, sellers and professionals. This technological evolution mirrors the revolutionary changes that the internet brought about in the late twentieth century, but it goes even further by streamlining processes, reducing errors and uncovering efficiencies that were previously unimaginable.

However, the shift towards an AI-driven market is not without its challenges. It necessitates a proactive approach from all stakeholders in the property market to adapt to new roles, to develop skills that complement AI and to foster a culture of continuous learning and adaptation. For professionals, embracing AI tools and technologies is not just a means to enhance their work but a necessary step to remain competitive and valuable in an increasingly digital market.

For buyers and sellers, the AI-powered property market promises a more informed, efficient and stress-free transaction process. The ability to access accurate, comprehensive and up-to-date information about properties and market conditions, coupled with streamlined communication and transaction processes, marks a significant improvement over the current system.

As we stand on the brink of the AI revolution in the property market, the potential for AI to redefine how we buy, sell and manage property is immense. By embracing this change, the property market can unlock new sources of value, improve the transaction experience for consumers and usher in a new era of efficiency and transparency. The journey towards an AI-powered future is both an exciting and a necessary evolution for the property market, promising to reshape the industry for the better.

CHAPTER 18

The propertyverse

> The next version of the internet may be a far more immersive virtual experience. While the concept of the 'Metaverse' has received renewed attention, many of its basic elements – like virtual and augmented reality, or cryptocurrency transactions – have been under construction for decades. By essentially making the internet a virtual twin of the physical world, this digital do-over could enable novel ways of working, buying things, learning, and socializing.
>
> — World Economic Forum[1]

At the beginning of this book, I explored the physical property market – the era before the internet, characterized by simplicity, certainty and security. Transactions were tangible, trust was inherent and the process was well understood by all parties involved.

I then went on to explore how the arrival of the internet and the online market saw the dematerialization of money, titles and identities and brought a whole new level of complexity and fragmentation to the property market. This has resulted in property professionals and consumers having to navigate a labyrinth of online portals, electronic documents and digital communication channels, often at the expense of efficiency and clarity. The once-simple process of transferring property ownership has become a convoluted web of disconnected technology, fragmented identities and siloed data, leaving many longing for the simplicity and cohesion of the past.

But what if we could combine the best of both worlds – the ease and simplicity of the physical property market with the power and potential of digital technology? Enter the metaverse, or as I call it, the propertyverse – a future environment for property transactions that promises to transform the industry by bringing back the simplicity and trust of the past while harnessing the capabilities of the future.

William Gibson, the well-known author and essayist who coined the term 'cyberspace', once said: 'The future is already here – it's just not very evenly distributed.' I think this is the case for the propertyverse. Many of the technological components necessary for its realization – such as a property market infrastructure, common data standards, trust frameworks powering trusted data, smart contracts, housing super apps, AI, and tokenized digital money, titles and identities – are either in development or already in existence. However, in the words of Gibson, these elements are not yet evenly distributed.

If you believe these capabilities will come to fruition, it is a small leap of the imagination to believe that they will converge – and if they do, we will have a more streamlined, transparent and user-friendly property market, in which the complexities and fragmentations introduced by the internet are replaced by a seamless, integrated ecosystem. We will have returned to the simplicity, certainty and security of the physical market but in the digital world. We will have arrived in the propertyverse.

The history of the metaverse

This journey into a more immersive digital world started in the 1960s when Morton Heilig created Sensorama, an early example of virtual reality (VR) technology. Sensorama was more than a simple experiment; it was a device designed to stimulate multiple senses, showing the potential of technology to mimic or even improve on real-life experiences by using sights, sounds, smells and touch. This invention demonstrated the potential for creating engaging digital environments that could match the richness of the real world.

Around the same time, Ivan Sutherland developed the Sword of Damocles, the first head-mounted VR display. Although basic compared with today's technology, it was a revolutionary step forward, introducing the idea of enhancing the physical world with digital images. This invention laid the groundwork for augmented reality (AR), which blends digital elements seamlessly into our daily lives.

Through the 1970s and 1980s VR technology continued to evolve, moving beyond novelty uses into training, simulation and entertainment. It was during this period that William Gibson, in his novel *Neuromancer*, coined the term 'cyberspace', envisioning a digital universe that users could navigate. This idea anticipated the concept of the 'metaverse' – a term introduced in Neal Stephenson's *Snow Crash* in 1992. Stephenson's metaverse described a vast virtual reality space, and the concept has grown to mean a network of virtual spaces where people can meet, work and play without the limitations of physical distance.

The 1990s saw the rise of virtual communities such as Second Life, where users could create, navigate and interact within digital worlds. This era showed the potential of VR to enable new kinds of social interaction and community building online. With the arrival of the twenty-first century, AR began to merge digital and physical worlds more smoothly, especially with the adoption of smartphones. AR applications could overlay digital information on the real world, changing how we perceive our surroundings. The 2010s brought significant progress in VR and AR technologies, with investments leading to advanced VR headsets such as the Oculus Rift and to AR platforms such as ARKit and ARCore, making immersive digital experiences more accessible to the general public.

As we entered the 2020s the idea of the metaverse started to gain mainstream recognition, particularly with Facebook changing its name to Meta in 2021. This change signalled a strong commitment to creating a more connected, immersive digital world, blending the virtual and the real more closely than ever before.

This history, from the early days of VR to the current vision of the metaverse, showcases ongoing digital innovation and the desire to go beyond the physical limits of existence. We can see this move towards more immersive experiences in home buying when we look at services such as Matterport, which offers 3D tours.[2] But way beyond those basic immersive experiences, the metaverse will open up new ways for us to interact, do business and explore digitally. As we stand at the start of this new digital era, the metaverse promises to significantly impact various sectors, including the property market, redefining our digital and physical realities in ways we have yet to fully imagine.

Avatars: the first digital humans

The term 'avatar' has a long and interesting history, dating back thousands of years to Hindu mythology. In Sanskrit, 'avatar' means 'descent', and it was used to describe the incarnation of a Hindu deity in human or animal form.

One of the earliest known uses of avatars in digital technology was in the 1978 computer game MUD1 (short for 'multi-user dungeon'). In this text-based game, players could create characters and interact with others in a virtual world. While there were no graphics, players could use text to describe their character's appearance and actions.[3]

The first graphical avatars appeared in the early 1980s with the release of online games such as Habitat, a virtual world created by Lucasfilm for the Commodore 64. In Habitat players could create customized avatars and interact with others in a 2D graphical environment. Other games, such as Ultima IV and Gauntlet, also allowed players to create graphical representations of themselves.[4]

In the late 1980s and early 1990s the use of avatars became more widespread through the rise of online chat rooms and forums. In early internet communities such as The Well and AOL, users could create simple avatars to represent themselves in online discussions.

As technology advanced in the 1990s and 2000s, avatars became more sophisticated and three-dimensional, with games such as World of Warcraft and Second Life allowing players to create highly customized avatars with detailed appearances and animations. Today, avatars are used in a wide range of contexts beyond gaming, including social media, VR and online education.

In the 2000s social media platforms such as Facebook and Twitter introduced profile pictures, which could be seen as a simplified form of an avatar. These images allowed users to represent themselves visually in online communities, and they continue to be an important part of social media culture today.

The evolution of digital humans

More recently, advances in VR and AR technology have led to the development of even more sophisticated avatars. Today's avatars can be incredibly lifelike, with realistic facial expressions and body movements, leading some to term them digital humans.

Digital humans, at their core, are virtual representations of people. However, their evolution – powered by AI – has elevated them from mere visual representations to intelligent digital agents. These entities are capable of learning, thinking and interacting in a sophisticated manner that mirrors, and in some cases enhances, human behaviours, preferences and emotions. This leap from static avatars to dynamic, intelligent digital personas signifies a pivotal shift towards the creation of digital doppelgangers or unique digital beings, marking a significant milestone in the evolution of digital identity.

One example of a company that is already using digital humans to represent brands online is UneeQ.[5] These digital humans communicate with customers in real time, providing them with confidence in their purchases. It is not far-fetched to imagine a future in which buyers and sellers in property transactions also use digital humans or avatars to represent them.

Beyond housing super apps: enter the propertyverse

Imagine a world in which buying and selling a property is as simple and secure as a transaction in the physical world, but with the added benefits of digital technology. Welcome to the propertyverse, a revolutionary new environment that seamlessly blends the best aspects of the traditional property market with the potential of the digital market.

In the propertyverse, buyers, sellers, estate agents, conveyancers, mortgage lenders and even HM Land Registry officials come together in a secure, immersive and interconnected digital space. This virtual ecosystem replicates the simplicity and immediacy of physical property transactions while using cutting-edge technologies to enhance efficiency, transparency and security.

For buyers, stepping into the propertyverse is like entering a space where the boundaries between the digital and physical worlds blur. As the prospective buyers put on their VR headsets, they are presented with a beautifully rendered 3D model of a potential new home. With a movement of their hand, a holographic interface materializes, granting them instant access to a wealth of property details. Detailed floor plans, high-definition imagery and real-time sensor data about the home's energy efficiency and smart features come alive through their headset, providing a comprehensive understanding of the property.

Eager to explore further, the buyers navigate the virtual environment, and their digital avatars move through each room. As they examine the property, they notice interactive prompts that allow them to customize the space to their exact preferences.

From cabinet finishes to counter-top materials, they can seamlessly experiment with various design options to visualize how their personal touch will transform the home.

The buyers can also view the property's ownership and transaction history by accessing a secure data store with a comprehensive audit trail, including details on previous owners, sale prices and even any past renovations or maintenance records.

As these buyers continue to explore their potential new home, they notice the possibility of a loft conversion. At their request, the digital twin of the property transforms, rendering the attic space in stunning 3D detail. They can customize the design, adjust the layout and even obtain a detailed cost estimate and the projected value increase of doing the conversion.

Finally, the buyers want to hear the seller's perspective on local schools, so they summon the seller's virtual representation. In a natural and immersive conversation, they are able to ask questions, understand the seller's view on local schools and gain valuable insights that would typically only be available during a physical viewing.

Impressed by the property and its potential, the buyers decide to put in an offer. Their agent, represented by a personalized avatar, talks through the offer, answers their questions and then informs the seller.

For estate agents and conveyancers, the propertyverse will be transformative. With AI-assisted tools and automated workflows, these professionals will be able to provide their clients with unparalleled service and support. Estate agents can harness the power of data analytics to identify the most suitable properties for their clients, while conveyancers can use smart contracts to streamline the legal process and ensure secure, error-free transactions.

In the propertyverse estate agents and conveyancers can meet with clients virtually, using their digital avatars to provide personalized advice and guidance. These avatars serve as secure digital identities that ensure every interaction is authentic and traceable. Collaborative tools allow all parties involved in a transaction to communicate seamlessly, share documents and track progress in real time, fostering a sense of transparency and trust.

The integration of mortgage lenders and Land Registry officials into the transaction process plays a crucial role in the propertyverse, and the digital ecosystem ensures that this integration is seamless. With access to verified digital identities and secure, tamper-proof data, mortgage lenders can make swift, accurate decisions on mortgage applications. The integration with a Regulated Liability Network enables both lenders and the Land Registry to issue digital money and titles directly into the propertyverse. Smart contracts then automate the terms and conditions of property transactions, ensuring that all parties fulfil their obligations – from the transfer of funds to the exchange of property ownership – without the need for manual intervention. No more registration gap, just real-time title transfers.

The propertyverse tech stack

The metaverse is structured like a multilayered ecosystem, with each layer supporting a different aspect of the digital universe. Jon Radoff and Matthew Ball, experts on the subject, have outlined this structure to help us understand how various technologies come together to form the metaverse. Below, I explore these layers and how they relate to the propertyverse.

The first is the experience layer. This is where users directly interact with the digital property world. It is essentially the user interface of the propertyverse, providing immersive experiences that can range from property searches to transaction management and property completion. Companies such as Epic Games, with Fortnite, and Unity Technologies, with their development platform, are key players in this layer, offering the tools and platforms that create these rich digital experiences. We can imagine consumers entering the propertyverse through a housing super app tailored to property transactions.

Next, we have the discovery layer, which is all about helping users find where they need to go in the propertyverse in order to connect with the property and the professionals they need for their transaction. It is a crucial component for navigating the vastness of a digital property world in which hundreds of thousands of transactions are happening and thousands of professionals are present. Again, we can imagine consumers accessing this layer through their property-focused housing super app.

The creator-economy layer supports the individuals and businesses that build the propertyverse's content. It provides a set of tools and platforms that enable creators to design, develop and monetize their digital property experiences. This layer lowers the barriers to content creation, allowing for a wider range of voices and creativity within the propertyverse. This layer would be created by proptech companies and in-house software teams, the same way portals for partners and customers are developed today.

Spatial computing forms another layer, focusing on the integration of the digital and physical property worlds. It includes technologies that enable digital property objects to interact with the physical environment, such as AR apps that overlay digital property information onto the real world. This layer blurs the line between digital and physical, creating seamless experiences that enhance our interaction with properties. Spatial computing offers significant opportunities for estate agents in letting them present properties for sale and dynamically 'dress and stage' properties to visualize clients' needs and requirements.

The decentralization layer incorporates blockchain technology into the propertyverse, providing a secure and transparent method for managing digital property ownership, identity and transactions. This layer is crucial for establishing the economic foundation of the propertyverse, in which real-world assets such as money and titles reside. As I discussed earlier in the book when exploring the development of digital asset and tokenization platforms (chapter 9) along with the possibilities of digital money and titles (chapters 10 and 11), this layer will play a pivotal role in integrating physical properties into the propertyverse, enabling the exchange of real-world assets in the virtual world.

The human-interface layer includes the hardware and software that allow users to interact with the propertyverse. This covers a wide range of devices, from VR headsets and AR glasses to more futuristic technologies such as brain–computer interfaces. These technologies are key to making the propertyverse accessible and immersive.

At the heart of the propertyverse lies the infrastructure layer, which provides the essential technological backbone for the entire ecosystem. This layer is anchored by the shared property market infrastructure, a critical component that ensures the smooth operation of the digital property universe. The property market infrastructure encompasses a range of technologies, including cloud computing services, data centres and networking protocols. These components work together to support the vast amounts of data and interactions that occur within the propertyverse, enabling seamless communication and data exchange between all participants in the property market.

In essence, the infrastructure layer, with the property market infrastructure at its core, serves as the bedrock on which the entire propertyverse is built. It provides the necessary stability, connectivity and efficiency to support the complex web of interactions and exchanges that define the future of the property market.

Together, these layers will form the propertyverse, a complex but coherent digital ecosystem that will extend the possibilities of how we connect and interact to transact property in the future.

Conclusion

As we conclude our exploration of the propertyverse, it is understandable that some may view this concept with scepticism.

After all, the idea of a fully integrated, digital ecosystem for property transactions may seem like a distant dream. However, the reality is that the building blocks of the propertyverse are already in place, and the benefits of this digital revolution are too significant to ignore.

It is important to recognize that the internet, while revolutionary, has also fragmented and complicated the property market. The once-simple process of buying and selling property has become a labyrinth of online portals, electronic documents and disparate communication channels. The propertyverse, on the other hand, represents what the internet should have been from the start: a combination of the simplicity and cohesion of the physical world with the capabilities of the digital world.

Throughout this book, I have discussed the various technological components that will form the foundation of the propertyverse, such as a property market infrastructure, common data standards, trust frameworks, smart contracts, tokenized digital assets and AI. While these elements may not yet be fully integrated or evenly distributed, their existence and ongoing development serve as proof points that the propertyverse is not merely a concept but a tangible reality in the making.

Moreover, the potential benefits of the propertyverse are compelling. By creating a seamless, secure and efficient digital environment for property transactions, the propertyverse promises to streamline processes, reduce costs and enhance the overall experience for buyers, sellers and professionals alike. The use of avatars and digital humans will enable more immersive, personalized interactions, while the integration of blockchain technology and smart contracts will ensure the security and transparency of transactions.

The propertyverse also presents significant opportunities for innovation and growth within the property industry. By lowering barriers to entry and enabling new forms of collaboration, this digital ecosystem will foster the development of novel solutions and business models. The creation of a shared property market infrastructure will provide a level playing field for all participants, promoting competition and driving the industry forward.

As with any transformative change, the journey towards the propertyverse will not be without its challenges. There will be legal and ethical considerations to address, as well as the need for industry-wide collaboration and standardization. However, the potential rewards of this digital revolution far outweigh the obstacles.

Ultimately, the propertyverse represents a natural evolution of the property market in the digital age. By combining the best aspects of the physical world with the power of emerging technologies, we can create a more accessible, transparent and user-friendly environment for property transactions. The proof points are there, and the benefits are clear. It is up to us, as industry leaders and innovators, to work together to shape the future of the property market in the propertyverse. The propertyverse is not just a vision of the future; it is the digital property market we always should have had – one that restores simplicity and trust while harnessing the boundless potential of technology.

CHAPTER 19

Market supervision

> The Bank's supervision of FMIs [financial market infrastructures] contributes to its mission to promote the good of the people of the UK by maintaining monetary and financial stability. The Bank seeks to ensure that the FMIs it regulates reduce systemic risk by ... identifying and mitigating risks in the end-to-end process of making payments, clearing and settling securities transactions, and clearing derivatives trades.
>
> — Bank of England, 2021[1]

The property market in England and Wales is facing significant challenges, with a lack of accountability and supervision being a major concern. Currently, no single entity is responsible for overseeing the market, which has led to a range of issues that affect both businesses and consumers. This absence of a dedicated supervisory body has created an environment in which businesses can operate without adequate security measures and oversight, contributing to the market's overall inefficiency and leaving it vulnerable to various risks, including cyber threats and fraud.

As a result of the lack of supervision, businesses in the property market often face challenges such as disrupted systems and conveyancing delays. These issues not only affect the businesses themselves but also have a direct impact on consumers, who may experience frustration, financial losses and a lack of trust in the market. To address these challenges and transform the property market into a thriving, sustainable and efficient ecosystem, effective supervision is crucial.

A well-designed supervisory framework would help protect consumers from potential risks, foster innovation and growth within the market and ensure that businesses operate with the necessary security measures in place. The financial sector, particularly in the UK, has demonstrated the benefits of having a robust supervisory framework. The Bank of England's role in overseeing financial market infrastructures (FMIs) has helped ensure stability, resilience and innovation within the sector. By adopting a similar approach, appropriately scaled for the property market, we can create a safer, more efficient and innovative environment for all stakeholders.

This transformation could not only address the current issues faced by businesses and consumers but also pave the way for a world-leading digital property market. In the following sections I explore the lessons that can be learned from the financial markets, the steps needed to transform the property market and the potential benefits of implementing a comprehensive supervisory framework.

Lessons from financial markets

The financial sector in the UK offers valuable insights into the benefits of having a robust supervisory framework. The Bank of England (BoE) plays a crucial role in ensuring market stability and fostering innovation within the sector. By setting clear standards and expectations for operational resilience, the BoE helps FMIs maintain their robustness and their ability to operate effectively, even under stress.

These standards address critical aspects such as recovery capabilities, cybersecurity and business-continuity planning. To ensure compliance with these standards, the BoE conducts regular supervision and assessment of FMIs. This process involves a range of activities, including checking compliance with resilience standards, evaluating risk management practices and reviewing business-continuity plans through on-site visits, interviews and detailed analyses.

In addition to setting standards and conducting assessments, the BoE actively promotes testing and collaboration among FMIs. Regular testing and simulation exercises help FMIs evaluate their response and recovery capabilities in the face of various disruptions. These exercises play a vital role in identifying and addressing potential weaknesses in the system, ultimately strengthening the overall resilience of the financial sector.

Collaboration and information sharing between FMIs and with regulatory authorities are also key aspects of the BoE's approach. By fostering an environment in which best practices, incident learnings and threat intelligence are shared, the BoE helps create collective resilience within the financial system. This collaborative approach enables coordinated responses to potential threats and enhances the sector's ability to withstand and recover from disruptions.

Furthermore, the BoE has enforcement powers to ensure compliance with operational resilience requirements. If vulnerabilities or instances of non-compliance are identified, the BoE can mandate actions to address these issues, impose penalties or even revoke operating authorization in severe cases. These enforcement powers serve as a strong incentive for FMIs to adhere to the established standards and maintain a high level of operational resilience.

The BoE's comprehensive strategy for ensuring the operational resilience of FMIs encompasses setting standards, regularly monitoring compliance, promoting testing and collaboration, exercising enforcement powers and engaging in international cooperation. By adopting a similar approach in the property market, tailored to its specific needs and characteristics, we can create a more stable, efficient and innovative ecosystem that benefits all stakeholders.

The need for property market supervision

The UK property market, despite its immense value and importance to the nation's economy, currently operates without adequate system-level government oversight. This lack of supervision is particularly alarming considering that the property market holds £8 trillion of national wealth and facilitates billions of pounds in asset transactions annually. While HM Land Registry plays a crucial role within the market, it has explicitly stated that it is not accountable for the overall functioning or dysfunction of the market. This situation highlights a significant gap in accountability and supervision that parliament has yet to address.

The absence of a comprehensive supervisory framework leaves the property market vulnerable to various risks and inefficiencies. Without proper oversight, businesses can operate without the necessary security measures, leaving them exposed to cyber threats and other vulnerabilities. This lack of supervision also contributes to a fragmented and opaque market, in which information asymmetries and inconsistencies can hinder efficiency and erode trust among participants.

To ensure that the property market operates with the desired level of transparency, efficiency and stability, a robust supervisory framework is essential. Such a framework would not only protect the interests of those directly involved in property transactions but also safeguard the broader economic interests of the nation. By establishing clear standards, monitoring compliance and promoting best practices, a supervisory body can foster a more resilient and innovative property market that benefits all stakeholders.

To achieve this goal and modernize the property market into a truly digital property market, we must redefine the roles of key institutions such as the Land Registry. This redefinition involves expanding their focus beyond merely overseeing legal titles to encompass a broader range of responsibilities aimed at enhancing the market's resilience, performance, stability and capacity for innovation.

By assuming the role of a market supervisor, the Land Registry can play a pivotal role in setting and enforcing standards, monitoring market performance and promoting collaboration among participants. This expanded role would enable the Land Registry to proactively identify and address potential risks, ensure the integrity of property transactions and foster an environment conducive to innovation and growth.

In summary, the current lack of system-level government oversight in the UK property market exposes it to various risks and inefficiencies. To address these challenges and ensure the market's transparency, efficiency and stability, a comprehensive supervisory framework is necessary. Redefining the role of key institutions such as HM Land Registry to include market supervision is a crucial step towards achieving this goal and transitioning to a world-leading digital property market.

HM Land Registry as market supervisor

Assuming that the Land Registry would be the organization responsible for property market supervision seems logical, as it is challenging to find a more suitable entity for this task. The HM Land Registry Act 2002, along with the preparatory work undertaken with the Law Commission, suggests that the Land Registry was envisioned to take on a much more supervisory role in the market. Specifically, the Act empowers the lord chancellor to regulate transactions that can be conducted electronically, indicating a broader scope for the Land Registry's functions.

It is not a significant leap to imagine the Land Registry transitioning from being merely a guardian of the register to an organization responsible for supervising the market. Nor is it a substantial leap to start thinking about the Land Registry as a digital asset custodian instead of a record-keeper. This transition to a supervisor and digital asset custodian would involve several key activities that would not only redefine the Land Registry's role but also enhance its capacity to support a more stable and efficient digital property market.

To transition from being a record-keeper to a market supervisor, the Land Registry must undergo a fundamental shift in perspective. Currently, its board is mainly comprised of legal and retail experts focused on ensuring record security and providing retail-style online services. This setup contrasts sharply with the Bank of England's leadership, which is predominantly made up of economists. Economists are, by nature, systems thinkers. They assess risk, performance and function at a market or system-wide level. Lawyers, on the other hand, tend to concentrate on details, prioritizing legal precedents and the intricacies of clauses for risk management.

This difference leads to a crucial question: who is better suited to help the Land Registry achieve its aim of creating a 'world-leading property market as part of a thriving economy and a sustainable future'? Is it a team of lawyers adept at navigating legal complexities, or a team of economists who understand the broader economic system? For HM Land Registry to realize its vision, incorporating strong systems thinkers into its board is essential. This suggests that broadening the board's expertise to include more economists or individuals with a systemic perspective could be a key step towards achieving its goals.

The next transformation or transition is from being a record-keeper to acting as a digital asset custodian. As covered in chapter 11, this recommendation involves fundamentally reimagining property titles as digital assets, akin to equities in the financial markets. Such a transformation is not merely cosmetic but signifies a paradigm shift with the potential to unlock unprecedented opportunities for growth and value creation in the market.

The Land Registry can pave the way for a much more fluid and active market, moving from a maximum of four people on a title to as many as the public requires, and from costly and risky title transfers to a seamless, safe and certain process. This evolution would enable innovation around property investment and ownership and magnify the economic impact of the property market. It would demand that we view property not just as static physical assets but as fluid, tradeable digital assets.

To support this transformation, the Land Registry should follow the guidance of the report on 'Digital transformation and land administration' co-authored by the United Nations Economic Commission for Europe and ally with market leaders in new technologies such as blockchain and tokenization.[2] These technologies promise enhanced security, transparency and efficiency in recording property transactions, thereby facilitating faster transactions, reducing fraud and bolstering trust within the market.

Moving HM Land Registry under HM Treasury

Currently, the Land Registry is under the oversight of the Department for Levelling Up, Housing and Communities, a placement that aligns with its traditional role as a custodian of land records. However, considering the Land Registry's impact on the national economy and its stewardship of assets worth £8 trillion, a re-evaluation of this alignment is necessary. The need for increased liquidity and efficiency in property transactions demands a shift in perspective.

As explored in chapter 11, the UK is witnessing considerable innovation in digital assets, including the tokenization of securities and funds. This pioneering work is led by HM Treasury, an organization that inherently adopts a systemic approach to market and economic functions.

To realize the Land Registry's vision of leading the property market towards a thriving economy and a sustainable future, it is essential to consider placing it under HM Treasury's governance. Such a move would benefit the Land Registry with systemic thinking and access to the digital asset innovation, market supervision, and identity and security expertise found within HM Treasury.

Towards a digitally mature market

In reflecting on the transition from an industry that is fragmented and without supervision to one that is digitally mature, supervised and competitive, it is clear that there should be some rules and minimum standards associated with doing business in the systemically important property market, where the stakes are incredibly high. The importance of standards and supervision cannot be overstated. Over the last eighteen months, significant portions of our market have been crippled by cyber-attacks, underscoring the critical need for robust security measures and oversight.

The rise of AI-powered bad actors and the increased systemic vulnerabilities of a fully digital property market further emphasize the crucial point that participation in the market should be contingent on meeting a base level of market standards, set by the supervisory body. As we move towards a more digitally mature market, it is essential that we address the risks posed by cyber threats and malicious AI, ensuring that all participants adhere to strict security protocols and best practices.

However, in our efforts to establish standards and supervision, we must also be mindful of the importance of maintaining an open market. The property market thrives on competition, innovation and the free flow of ideas, and it is crucial that we strike a balance between regulation and openness. An open market should not equate to a dysfunctional one with no standards or rules, but neither should it be so heavily regulated that it stifles growth and creativity.

Finding this balance will require careful consideration and collaboration among all stakeholders, including regulators, industry leaders and technology providers. By working together to establish clear guidelines, promote best practices and foster a culture of continuous improvement, we can create a digitally mature property market that is both secure and dynamic.

The journey towards a digitally mature market will undoubtedly present challenges, but the potential benefits – in terms of economic growth, social well-being and technological progress – are too significant to ignore. As we navigate this transition, it is essential that we remain committed to the principles of transparency, accountability and innovation, ensuring that the property market serves the needs of all participants, from buyers and sellers to professionals and regulators.

Ultimately, the success of our efforts to create a digitally mature property market will depend on our ability to adapt to new technologies and work together towards a common goal. By doing so, we can not only address the current challenges facing the market but also lay the foundations for a more resilient, efficient and equitable future.

AI governance and the impact–autonomy framework

As AI continues to advance and permeate various sectors of the property market, it is crucial to consider the governance of this powerful technology. To effectively assess and manage the implications of AI in the property market, I propose the impact–autonomy (IA) framework. This two-dimensional matrix examines AI's current and prospective effects along two axes: the breadth of socio-economic impact and the degree of autonomous decision making.

By charting AI use cases along these axes, stakeholders can derive valuable insights into the appropriate development pacing, control mechanisms and policies to achieve beneficial innovation while managing risks.

The socio-economic impact dimension

The socio-economic impact dimension of the IA framework considers how AI will affect different levels of the economy and society. This dimension is divided into four levels.

- *Individual.* At this level, AI enhances human potential and productivity without replacing core functions. It assists professionals in the property market by automating routine tasks, providing insights and enabling them to focus on higher-value activities. Risks are limited as long as individuals understand that AI augments rather than replaces their capabilities.

- *Organizational.* AI targets internal operational efficiency gains at this level, taking on routine analytical and interactive tasks. It can help property companies streamline processes, improve decision making and enhance customer experiences. Humans provide oversight and focus on higher-order work such as strategy, innovation and stakeholder relations.

- *Market.* Here, AI catalyses the emergence of new data-centric business models and transforms sectors' economics through precise predictions and automation. It can enable the creation of new property-related services, improve market transparency and facilitate more efficient transactions. However, inequality may widen if technology consolidation accrues more gains to capital than to labour.

- *Nation-state.* At this level, AI improves the broad quality of life by transforming government services. In relation to property, these would include urban planning, infrastructure development and housing policy. It can help governments make more informed decisions, optimize resource allocation and better serve citizens. However, the risks range from digital authoritarianism to autonomous decision-making systems that lack human ethical judgment.

The autonomous decision-making dimension

The autonomous decision-making dimension of the IA framework evaluates the degree of independence granted to AI in making decisions. This dimension is also divided into four levels, inspired by the RACI matrix (responsible, accountable, consulted, informed).

- *Humans accountable, humans responsible, AI consulted.* At this stage, humans remain fully accountable and responsible for decisions, consulting AI to gather informational insights as an additional input for consideration. AI systems provide recommendations and support, but humans make the final call.

- *Humans accountable, AI responsible.* Humans set predefined constraints and objectives, enabling AI to take on responsibilities for key repetitive decisions. For example, AI might be responsible for automating certain aspects of the property search or transaction process based on predefined criteria. However, human accountability persists throughout the whole system and its monitoring procedures.

- *AI accountable, AI responsible, humans consulted.* AI owns accountability for autonomous decisions while integrating human domain perspectives as additional signals to enrich understanding. Humans advise without authority over outcomes. This level might involve AI systems independently conducting property valuations or risk assessments, with humans providing input but not overriding the AI's decisions.

- *AI accountable, AI responsible (fully autonomous).* AI has full responsibility to independently make consequential decisions based on core objective criteria, without external overrides or limiting guard-rails. Humans can only review system-wide patterns retrospectively and enact policy restrictions on overall activity. This level is not yet feasible or desirable for most property market applications, as it would require AI systems to operate with a level of autonomy and ethical reasoning that is currently beyond our capabilities.

Using the framework to assess risks

The IA framework provides a structured approach for assessing AI's current and future effects by considering both the socio-economic impact and the freedom of autonomous decision making. This enables stakeholders in the property market to develop strategies that maximize AI's benefits while minimizing potential risks and negative consequences.

By plotting AI initiatives on the IA axes, stakeholders can identify the appropriate level of human oversight, governance mechanisms and risk mitigation strategies required for each use case. For example, AI projects with a high socio-economic impact and high autonomy may require more stringent governance, transparency and accountability measures than those with a lower impact and lower autonomy.

Furthermore, the IA framework can help organizations anticipate and plan for the potential trajectory of their AI initiatives as they evolve along the impact and autonomy dimensions. This foresight allows stakeholders to proactively adapt their strategies, ensuring that AI development aligns with organizational values, market dynamics and regulatory requirements.

Stakeholders can use the IA framework to foster a culture of responsible AI innovation, so that AI systems are designed and deployed in a way that is ethical, transparent and accountable. This involves engaging in ongoing dialogue with internal and external stakeholders, monitoring AI systems for unintended consequences and being prepared to adjust course as needed.

Ultimately, the goal is to harness the transformative potential of AI to create a more efficient, accessible and equitable property market while mitigating risks and promoting the well-being of individuals, organizations and society as a whole. The IA framework provides a valuable tool for navigating this complex landscape and making informed decisions about the development and deployment of AI technologies in the property market.

Conclusion

The property market in England and Wales faces significant challenges due to the lack of a comprehensive supervisory framework. The absence of accountability and oversight has led to various issues, including cyber threats, conveyancing delays and a general lack of trust in the market. To address these challenges and transform the property market into a thriving, sustainable and efficient ecosystem, it is crucial to establish effective supervision and governance.

Drawing lessons from the financial sector, where the Bank of England plays a vital role in ensuring market stability and fostering innovation, the property market can benefit from adopting a similar approach. By setting clear standards, promoting collaboration among stakeholders and redefining the role of key institutions such as the Land Registry to include market supervision, we can create a more transparent, resilient and innovative property market.

The transformation of the Land Registry from a mere record-keeper to a market supervisor and digital asset custodian is a critical step in this journey. This transition involves moving the Land Registry from the Department for Levelling Up, Housing and Communities to under HM Treasury, incorporating systems thinkers into leadership positions, reimagining property titles as dynamic digital assets and using technology for enhanced security, transparency and efficiency.

As the property market moves towards digital maturity, it is essential to establish rules and minimum standards for participation, address the risks posed by cyber threats and malicious AI, and strike a balance between maintaining an open market and ensuring adequate supervision. The journey towards a digitally mature market requires collaboration among regulators, industry leaders and technology providers to establish clear guidelines, promote best practices and foster a culture of continuous improvement.

Furthermore, the impact–autonomy framework provides a valuable tool for assessing and managing the implications of AI in the property market. By considering the potential impact of AI at various levels (individual, company, market and nation-state) and the degree of autonomy granted to AI systems, stakeholders can develop strategies that maximize the benefits of AI while mitigating risks and promoting responsible innovation.

Ultimately, the success of creating a world-leading digital property market depends on the collective efforts of all stakeholders to embrace change, adapt to new technologies and work together towards a common goal. By doing so, we can not only address the current challenges facing the market but also lay the foundations for a more resilient, efficient and equitable future, potentially paving the way for the world's first regulated propertyverse.

Epilogue

> This I do know: at such times, it is no failure to fall short of realizing all that we might dream – the failure is to fall short of dreaming all that we might realize. We must try!
>
> — Dee Hock, founder of Visa

As we reach the end of our exploration of the property market, I sincerely hope that this journey has been as illuminating for you as it has been for me throughout the research and writing process. I have shared insights and observations, knowing well that the future may prove them right or challenge them. It is only natural for us to agree on some points and disagree on others, but I trust that any debates sparked by my critiques or perspectives are taken in the spirit of constructive discourse.

The property market is a complex and intricate web, and while it may be simpler to critique, my intention has always been to engage with these issues sincerely and transparently, offering solutions to the market's persistent and stubborn challenges.

This book serves as a call to action, urging us to move beyond just discussing the digital property market and instead focus on actively building the missing digital foundations that it so desperately needs. From there, we have a tremendous opportunity to embrace digital assets, AI and immersive technologies to create a next-generation intelligent property market. This market would not only solve today's problems but also meet the future aspirations of both industry professionals and consumers.

The UK is leading the world with innovative initiatives such as the Digital Securities Sandbox and the Regulated Liability Network. We can further cement our position as a global leader by pioneering digital titles and property assets. However, to do so, we need to shift our perspective and see property assets in a new light. We must design custody and transaction services that are fit for the £8 trillion worth of property assets entrusted to HM Land Registry.

We stand on the cusp of a wave of innovation so profound that it promises to redefine our very conceptions of property, ownership and value exchange. The propertyverse heralds a future in which digital assets, smart contracts and blockchain technology merge the physical with the virtual, ushering in an era of unprecedented accessibility, transparency and connectivity.

As we look forward to the road ahead, I offer a note of caution: the wave of change coming may be too profound for us to fully prepare for in time. That is why I believe that systems thinking and systemic monitoring and supervision are critical. What is certain is that the machines are coming, and unlike the adoption of the internet, we will not have the luxury of adopting them at our own pace and ignoring them in the meantime.

I hope this book has advanced the conversation around the digital property market, shedding light on what it is, what it can be, the missing pieces and what we must do to seize the opportunity.

Thank you for reading, and please do not hesitate to reach out and connect digitally.

Acknowledgements

I am deeply grateful to the many individuals whose wisdom, critique and insights have profoundly influenced my understanding of the property market. This book stands as a testament to countless discussions with collaborators who have not only enriched my perspectives but also inspired actionable insights. It is my sincere hope that their generous sharing of knowledge and vision is accurately reflected and paid forward in these pages.

I want to start by acknowledging the work, friendship and partnership of my co-founders at Coadjute, Dan Salmons and Sanj Bulsara. To Dan, I extend my gratitude for your deep intellectual challenge and thought partnership, which has been pivotal in clarifying and deepening my understanding of the market's complexities. And to Sanj, I am grateful for your tireless efforts in turning our product vision into reality.

To the incredible team at Coadjute, your persistence, grit, perseverance and relentless dedication to advancing the property market have been truly inspiring.

However, this journey extends well beyond Coadjute's boundaries. Our partners, customers and stakeholders have played a crucial role in expanding my horizons and providing perspectives that have been essential to my learning. While it is impossible to name everyone, I must acknowledge those who have stood out as thought partners and challengers, offering insights that have significantly influenced my thinking: Andrew Asaam, Andy Barrow, Annie Birchall, Bethany De Montjoie Rudolf, Richard Brown, Richard G. Brown, Kate Buckland, Libby Chambers, Claire Cherrington, Isabelle S. Corbett, Pauline Cowan, Dave

Cray, Eddie Davies, Edward Norfolk, Steve Dawkins, Mike Day, Esther Dijkstra, Matthew Fernandes, David Foreman, Glynis Frew, Professor Sarah Green, Paolo Guida, Chris Hackworth, Rob Hailstone, Mike Harlow, Maria Harris, Sally Holdway, John Jackson, Henry Jordan, Andrew Knight, Diane Later, Peter Left, Tara Lourens, Anish Malhotra, Mark Manning, Stephen McKeon, Tony McLaughlin, Ed Molyneux, Nisha Morjaria, Dovile Naktinyte, Kate Faulkner OBE, Professor Stewart Brymer OBE, Jurin Ohyama, Richard Price, Francesca Hopwood Road, Charlotte Sadd, Miguel Sard, Miles Shipside, James Slater, Imran Soomro, Matt Spence, Aime Stokes, Johan Svanstrom, Lauren Tombs, Rees Watkins, Lydia Yao and Heinrich Zetlmayer.

Every conversation and challenge has been instrumental in my quest to understand and contribute to the evolution of the property market. This book seeks to capture and broaden the collective wisdom and progressive spirit of all who have journeyed with me. I extend my deepest thanks for your steadfast support, insight and inspiration.

A special acknowledgement goes to John Abbott for your vision in establishing the Digital Street project, which set me on this path; to David Rutter for supporting me and Coadjute from the outset; and to David Birch, whose books have inspired me. A chance meeting in London with David motivated me to make one final push to complete this book.

My gratitude also goes to Richard Baggaley and Sam Clark of London Publishing Partnership for having faith in me as a first-time author and for believing this book deserved to be published; to Diane Coyle for her wonderful work in editing the book; and to Alex Chambers of T&T Productions for his expertise in creating the final production script.

Finally, my thanks go to Dee Hock, Visa's founder and CEO emeritus who died in 2022, and whose spirit and wisdom flow through much of this book.

So to everyone mentioned, and many more, your support, insight and inspiration have been invaluable. Thank you.

About the author

John Reynolds is co-founder and chief operating officer of Coadjute, the award-winning SaaS platform and property market infrastructure backed by Rightmove, Lloyds Banking Group, Nationwide and NatWest.

Prior to entering the property market, John spent twenty-five years leading large-scale digital transformation initiatives across the UK private and public sectors.

In 2018 John partnered with enterprise blockchain firm R3 to lead and win a bid for HM Land Registry's Digital Street project, where, in collaboration with the Land Registry's Digital Street Community, his consultancy business built a prototype digital backbone for the property market, demonstrating the value of tokenized titles and a UK property market infrastructure. This project laid the foundation for the creation of the business that is today Coadjute.

Since Coadjute's inception, John has led the technology development of the Coadjute network, which is built using enterprise distributed ledger technology. This has given John a unique view of the digital workings of the property market across sectors, from estate agency and brokerage to lending and conveyancing, as well as consumer services.

John's contributions extend beyond Coadjute – he was one of the founding architects behind the Home Buying and Selling Group's Property Trust Framework and data standards, and a founding member of the UK's Open Property Data Association.

In 2022 John led Coadjute's bid to work with the Bank of England, the Bank for International Settlements and HM Land

Registry on Project Meridian, and he subsequently led the development of the prototype synchronization service, which demonstrated the real-time movement of funds in the RTGS and updates to the land register. Building on the success of Project Meridian, John has led Coadjute's work with Mastercard and the Regulated Liability Network and is now heading Coadjute's AI Labs, exploring the integration of artificial intelligence to further transform the property market.

This book is an invitation to see the property market through the eyes of someone who has been at the forefront of its digital evolution. John's insights offer a unique blend of practical experience and visionary thinking, making this book an indispensable guide for anyone interested in the intersection of technology, money and property. Whether you are a professional in the field, a policymaker or simply curious about the future of property transactions, John's story and insights provide compelling reasons to explore what lies ahead for the digital property market.

Endnotes

In compiling the citations for this book, I have endeavoured to include all the principal sources that have significantly informed its content. Given the journey of over seven years in the making of this work, it is conceivable that some influences and insights have been integrated into the narrative without explicit individual acknowledgement. These omissions are not for want of gratitude but rather an inadvertent consequence of the vast array of knowledge encountered over such an extended period.

I extend my deepest appreciation to all those whose ideas and wisdom have contributed to shaping my understanding, even if their specific contributions are not individually cited. This book is a testament to the collective knowledge and inspiration that have guided me through its creation.

For those interested in exploring the topics discussed further, the following notes detail key readings and sources that have been pivotal. I invite readers seeking a deeper engagement with any subject matter to connect with me on LinkedIn, where I am always eager to discuss ideas and share insights.

Preface

1. HM Land Registry. 2023. New Digital Property Market Steering Group formed to drive crucial digital transformation in the land and property market. News article, 1 August (www.gov.uk/government/news/new-digital-property-market-steering-group-formed-to-drive-crucial-digital-transformation-in-the-land-and-property-market).
2. HM Land Registry. 2022. HM Land Registry Strategy 2022+: enabling a world-leading property market. Report, 31 August (www.gov.uk/government/publications/hm-land-registry-strategy-2022).

Chapter 2

1. See www.parliament.uk/about/living-heritage/transformingsociety/towncountry/towns/overview/newtowns/.
2. Ben Jones. 2010. Slum clearance, privatization, and residualization: the practices and politics of council housing in mid-twentieth-century England. *Twentieth Century British History* **21**(4), 510–539 (https://ueaeprints.uea.ac.uk/id/eprint/40440/1/Slum_clearance_20th_Century_British_History_Vol_21_no_4.pdf).
3. Anna Minton. 2015. Byker Wall: Newcastle's noble failure of an estate – a history of cities in 50 buildings, day 41. *The Guardian*, 21 May (www.theguardian.com/cities/2015/may/21/byker-wall-newcastles-noble-failure-of-an-estate-a-history-of-cities-in-50-buildings-day-41).
4. Academy of Urbanism. 2018. From concrete to glass: the post-war trajectory of London's high-rise housing. Blog post, 17 January (www.academyofurbanism.org.uk/from-concrete-to-glass-the-post-war-trajectory-of-londons-high-rise-housing/).
5. Frank Eardley. 2022. Right to buy: past, present and future. House of Lords Library, 17 June (https://lordsli brary.parliament.uk/right-to-buy-past-present-and-future/).
6. Rowan Moore. 2023. From right to buy to housing crisis: how home ownership killed Britain's property dream. *The Guardian*, 29 October (www.theguardian.com/society/2023/oct/29/right-to-buy-housing-crisis-home-ownership-britain-property-rowan-moore).
7. Alex Bowen, Glenn Hoggarth and Darren Pain. 1999. The recent evolution of the UK banking industry and some implications for financial stability. Conference paper, 18 March, Bank for International Settlements (www.bis.org/publ/confp07l.pdf).
8. See part II of the Act: www.legislation.g ov.uk/ukpga/1985/61/part/II.
9. UK Parliament. 2024. Levelling Up Committee launches inquiry on improving the home buying and selling process. News article, 26 March (https://committees.parliament.uk/work/8373/improving-the-home-buying-and-selling-process/news/200646/levelling-up-committee-launches-inquiry-on-improving-the-home-buying-and-selling-process/).
10. Barclays. 2017. From the archives: the ATM is 50. News article, 27 June (https://home.barclays/news/2017/06/from-the-archives-the-atm-is-50/).
11. Sanjeev Kumar. 2021. Back to the future: the evolution of branchless banking. Blog post, 6 October, WhiteSight (https://whitesight.net/back-to-the-future-the-evolution-of-branchless-banking/).

Chapter 3

1. Paul Penrose. 2007. Who launched the UK's first Internet bank? *Finextra*, 23 May (www.finextra.com/blogposting/237/who-launched-the-uks-first-internet-bank).
2. Brian McCullough. 2018. A revealing look at the dot-com bubble of 2000 – and how it shapes our lives today. TED Ideas, 4 December (https://ideas.ted.com/an-eye-opening-look-at-the-dot-com-bubble-of-2000-and-how-it-shapes-our-lives-today/).
3. See https://plc.rightmove.co.uk/our-history/.
4. Tim O'Reilly. 2005. What is Web 2.0? Design patterns and business models for the next generation of software. O'Reilly Media, 30 September (www.oreilly.com/pub/a/web2/archive/what-is-web-20.html#mememap).
5. Reena Sewraz. 2023. Online estate agents. *Which?*, 17 January (www.which.co.uk/money/mortgages-and-property/home-movers/selling-a-house/online-estate-agents-a1gL85b08oxw).
6. UK Parliament. Levelling Up Committee launches inquiry. (See note 9 in chapter 2.)
7. See www.bankofengland.co.uk/news/2018/september/the-financial-crisis-ten-years-on.
8. Åke Grönlund and Thomas A. Horan. 2005. Introducing e-Gov: history, definitions, and issues. *Communications of the Association for Information Systems* **15**, 713–729 (https://doi.org/10.17705/1CAIS.01539).
9. Law Commission. 2001. Land registration for the twenty-first century: a conveyancing revolution. Report, 9 July (https://lawcom.gov.uk/project/land-registration-for-the-21st-century/).
10. See www.birketts.co.uk/legal-update/what-is-the-registration-gap/.
11. HM Land Registry. 2006. Land Registry annual report and accounts 2005/6. July (https://assets.publishing.service.gov.uk/government/uploads/system/uploads/attachment_data/file/231599/1434.pdf).
12. HM Land Registry. 2008. Land Registry annual report and accounts 2007/8. July (https://assets.publishing.service.gov.uk/media/5a7c8ec2ed915d6969f45b82/0767.pdf).

Chapter 4

1. HM Land Registry. HM Land Registry Strategy 2022+. (See note 2 in preface.)

2   Ministry of Housing, Communities and Local Government. 2018. Improving the home buying and selling process: summary of responses to the Call for Evidence and government response. Report, April (https://assets.publishing.service.gov.uk/media/5ac77d44e5274a5adc960e49/Improving_the_home_buying_and_selling_process_response.pdf).
3   Hannah Cromarty. 2022. Improving the home buying and selling process in England. Research briefing, 3 May, House of Commons Library (https://researchbriefings.files.parliament.uk/documents/SN06980/SN06980.pdf).
4   Conveyancing Association. 2016. Modernising the home moving process. White paper, November (www.conveyancingassociation.org.uk/wp-content/uploads/2023/11/Modernising-the-Home-Moving-Process-White-Paper-UPDATED.pdf).
5   UK Parliament. Levelling Up Committee launches inquiry. (See note 9 in chapter 2.)
6   Cromarty. Improving the home buying and selling process in England.
7   See www.birketts.co.uk/legal-update/what-is-the-registration-gap/.
8   Chris Pope. 2021. How do you solve a problem like requisitions? Blog post, 24 February, HM Land Registry (https://hmlandregistry.blog.gov.uk/2021/02/24/how-do-you-solve-a-problem-like-requisitions/).

Chapter 5

1   See www.legislation.gov.uk/ukpga/2002/9/section/92/notes.
2   Kelly Steed. 2023. UK mortgage statistics, 2023. Uswitch, 7 June (www.uswitch.com/mortgages/mortgage-statistics/).
3   Law Commission. Land registration for the twenty-first century: a conveyancing revolution. (See note 9 in chapter 3.)
4   Susan V. Scott and Markos Zachariadis. 2014. *The Society for Worldwide Interbank Financial Telecommunication (SWIFT): Cooperative Governance for Network Innovation, Standards, and Community*. Routledge.
5   See www.legislation.gov.uk/ukpga/2002/9/notes/division/2/4.
6   HM Land Registry. HM Land Registry Strategy 2022+. (See note 2 in preface.)
7   Lauren Tombs. 2019. Could blockchain be the future of the property market? Blog post, 24 May, HM Land Registry (www.digitalmarketplace.service.gov.uk/digital-outcomes-and-specialists/opportunities/6621).

8 John Reynolds and Dan Salmons. 2021. Coadjute: end-to-end property market infrastructure on Corda. Video presentation, 4 October, R3 (https://r3.com/videos/coadjute-end-to-end-property-market-infrastructure-on-corda/).
9 John Reynolds. 2023. Orchestration: the digital backbone of the property market. Blog post, 12 December, Coadjute (https://coadjute.com/blogs/orchestration-the-digital-backbone-of-the-property-market/).

Chapter 6

1 Department for Digital, Culture, Media and Sport. 2020. National Data Strategy. Policy paper, 9 December (www.gov.uk/government/publications/uk-national-data-strategy/national-data-strategy#executive-summary).
2 Central Digital and Data Office. 2023. Data Maturity Assessment for Government. Cabinet Office, 27 March (www.gov.uk/government/collections/data-maturity-assessment-for-government).
3 Geospatial Commission. 2022. How FAIR are the UK's national geospatial data assets? Assessment of the UK's national geospatial data assets. Policy paper, 9 February (www.gov.uk/government/publications/how-fair-are-the-uks-geospatial-assets/how-fair-are-our-national-geospatial-data-assets-assessment-of-the-uks-national-geospatial-data-html).
4 Swift. 2020. ISO 20022: the payments data revolution. News article, 24 September (www.swift.com/news-events/news/iso-20022-payments-data-revolution).
5 Department for Digital, Culture, Media and Sport. National Data Strategy.
6 See https://bills.parliament.uk/bills/3430/publications.
7 See www.gov.uk/government/groups/smart-data-working-group.
8 Department for Business and Trade. 2023. The Smart Data Roadmap: action the government is taking in 2024 to 2025. Report, April (https://assets.publishing.service.gov.uk/media/66190f98679e9c8d921dfe44/smart-data-roadmap-action-the-government-is-taking-in-2024-to-2025.pdf).
9 PwC. 2019. 'Putting a value on data'. Report (www.pwc.co.uk/data-analytics/documents/putting-value-on-data.pdf).
10 See https://homebuyingandsellinggroup.co.uk/about/.
11 See https://homebuyingandsellinggroup.co.uk/download/baspi-3/.

12 See https://github.com/Property-Data-Trust-Framework.
13 See https://openpropdata.org.uk/.
14 See www.lawsociety.org.uk/topics/property/transaction-forms.

Chapter 7

1 Building Regulations Advisory Committee. 2021. Building Regulations Advisory Committee: golden thread report. Policy paper, 21 July, Department for Levelling Up, Housing and Communities (www.gov.uk/government/publications/building-regulations-advisory-committee-golden-thread-report).
2 Independent Review of Building Regulations and Fire Safety. 2018. Building a safer future. Report, May, Ministry of Housing, Communities and Local Government (www.gov.uk/government/publications/independent-review-of-building-regulations-and-fire-safety-final-report).
3 These three challenges – identity confirmation, the data–identity link and the transfer of data – are being addressed by the Smart Data Working Group (www.gov.uk/government/groups/smart-data-working-group), the Open Property Data Association (https://github.com/Property-Data-Trust-Framework) and the Home Buying and Selling Group (https://homebuyingandsellinggroup.co.uk/download/baspi-3/).
4 Kim Cameron. 2005. The laws of identity. Paper published on *Kim Cameron's Identity Weblog* (www.identityblog.com/stories/2005/05/13/TheLawsOfIdentity.pdf).
5 David Birch and Victoria Richardson. 2023. *Money in the Metaverse: Digital Assets, Online Identities, Spatial Computing and Why Virtual Worlds Mean Real Business*. London Publishing Partnership.
6 Building Regulations Advisory Committee. Golden thread report.
7 Department for Science, Innovation and Technology, Department for Digital, Culture, Media and Sport and Julia Lopez MP. 2022. UK digital identity and attributes trust framework: beta version. Policy paper, 13 June (www.gov.uk/government/publications/uk-digital-identity-and-attributes-trust-framework-beta-version).
8 See https://bills.parliament.uk/bills/3430/publications.
9 See https://trustoverip.org/about/.
10 Department for Levelling Up, Housing and Communities. 2022. Levelling up the United Kingdom. Policy paper, 2 February (www.gov.uk/government/publications/levelling-up-the-united-kingdom).

11 National Trading Standards Estate and Letting Agency Team. 2023. Material information in property listings (sales). Version 1.0, November (www.nationaltradingstandards.uk/uploads/Material%20 Information%20in%20Property%20Listings%20(Sales)%20v1.0.pdf).
12 See www.lawsociety.org.uk/topics/property/transaction-forms.

Chapter 8

1 Margaret Rouse. 2021. Web 1.0. *Techopedia* (www.techopedia.com/definition/27960/web-10).
2 O'Reilly. What is Web 2.0? (See note 4 in chapter 3.)
3 Glenn Fleishman. 2000. Cartoon captures spirit of the internet. *New York Times*, 14 December (www.nytimes.com/2000/12/14/technology/cartoon-captures-spirit-of-the-internet.html).
4 David Birch. 2014. *Identity Is the New Money*. London Publishing Partnership.
5 Rouse. Web 1.0.
6 O'Reilly. What is Web 2.0?
7 Evan Karnoupakis. 2023. *NFTs, the Metaverse, and Everything Web 3.0*. O'Reilly Media.
8 Department for Science, Innovation and Technology et al. UK digital identity and attributes trust framework. (See note 7 in chapter 7.)
9 HM Land Registry. 2023. Improving today, building for tomorrow. Blog post, 28 July (https://hmlandregistry.blog.gov.uk/2023/07/28/improving-today-building-for-tomorrow/).
10 Scott and Zachariadis. *The Society for Worldwide Interbank Financial Telecommunication*. (See note 4 in chapter 5.)
11 See https://bills.parliament.uk/bills/3430/publications.
12 Eric Schmidt and Jared Cohen. 2013. *The New Digital Age: Reshaping the Future of People, Nations and Business*. Alfred A. Knopf.
13 Birch. *Identity Is the New Money*.
14 Department for Science, Innovation and Technology et al. UK digital identity and attributes trust framework.
15 World Economic Forum. 2020. Reimagining digital identity: a strategic imperative. Report, January (https://canada-ca.github.io/PCTF-CCP/docs/RelatedPolicies/WEF_Digital_Identity.pdf).
16 Department for Science, Innovation and Technology et al. UK digital identity and attributes trust framework.

17 See www.gov.uk/government/publications/identity-proofing-and-verification-of-an-individual#full-publication-update-history.
18 World Wide Web Consortium. 2022. Verifiable credentials data model v1.1. W3C recommendation, 3 March (www.w3.org/TR/vc-data-model/).
19 Alex Preukschat and Drummond Reed. 2021. *Self-Sovereign Identity: Decentralized Digital Identity and Verifiable Credentials.* Manning.

Chapter 9

1 Frank L. Holt. 2021. The invention of the first coinage in ancient Lydia. *World History Encyclopedia* (www.worldhistory.org/article/1793/the-invention-of-the-first-coinage-in-ancient-lydi/).
2 Tim Harford. 2017. What tally sticks tell us about how money works. *BBC*, 10 July (www.bbc.co.uk/news/business-40189959).
3 See www.historyskills.com/classroom/ancient-history/anc-gallic-wars-reading/.
4 Lawrence Chard. 2018. A brief history of coinage in Britain. Chards, 3 December (www.chards.co.uk/guides/brief-history-of-british-coins/464).
5 See www.royalmint.com/globalassets/the-royal-mint/pdf/history-of-the-royal-mint.pdf.
6 Bank for International Settlements. 2023. Blueprint for the future monetary system: improving the old, enabling the new. In *Annual Economic Report: June 2023*, chapter 3 (www.bis.org/publ/arpdf/ar2023e3.pdf).
7 Satoshi Nakamoto. 2008. Bitcoin: a peer-to-peer electronic cash system. White paper (http://dx.doi.org/10.2139/ssrn.3440802).
8 Vitalik Buterin. 2014. Ethereum white paper: a next generation smart contract and decentralized application platform. December (https://ethereum.org/content/whitepaper/whitepaper-pdf/Ethereum_Whitepaper_-_Buterin_2014.pdf).
9 Token Taxonomy Initiative. 2019. Moving tokens forward. Slide presentation (https://wiki.hyperledger.org/download/attachments/24775128/TTF-Presentation.pdf).
10 Tony McLaughlin. 2021. The regulated internet of value. Report, Citi (www.citibank.com/tts/sa/insights/articles/article191.html).
11 Sean Foley, Jonathan R. Karlsen and Tālis J. Putniņš. 2019. Sex, drugs, and bitcoin: how much illegal activity is financed through cryptocurrencies? *Review of Financial Studies* **32**(5), 1798–1853 (https://doi.org/10.1093/rfs/hhz015).

Chapter 10

1. Bank for International Settlements. Blueprint for the future monetary system. (See note 6 in chapter 9.)
2. Andrew Bailey. 2023. New prospects for money. Speech, 10 July, Bank of England (www.bankofengland.co.uk/speech/2023/july/andrew-bailey-speech-at-the-financial-and-professional-services-dinner).
3. See www.weforum.org/organizations/ripple/.
4. Bank of England. 2014. The economics of digital currencies – Quarterly Bulletin. YouTube video, 11 September (https://youtu.be/rGNNiTaC2xs?si=J3IAMJCs7BAUejzm).
5. Robleh Ali, John Barrdear, Roger Clews and James Southgate. 2014. Innovations in payment technologies and the emergence of digital currencies. In *Quarterly Bulletin: 2014 Q3*, Bank of England (www.bankofengland.co.uk/quarterly-bulletin/2014/q3/innovations-in-payment-technologies-and-the-emergence-of-digital-currencies).
6. Ben Broadbent. 2016. Central banks and digital currencies. Speech, 2 March, Bank of England (www.bankofengland.co.uk/speech/2016/central-banks-and-digital-currencies).
7. Mark Carney. 2018. The future of money. Speech, 2 March, Bank of England (www.bankofengland.co.uk/speech/2018/mark-carney-speech-to-the-inaugural-scottish-economics-conference).
8. Bank of England. 2020. Central Bank Digital Currency: opportunities, challenges and design. Discussion paper, 12 March (www.bankofengland.co.uk/paper/2020/central-bank-digital-currency-opportunities-challenges-and-design-discussion-paper).
9. Bank of England and Bank for International Settlements Innovation Hub. 2023. Project Rosalind: developing prototypes for an application programming interface to distribute retail CBDC. Report, June (www.bis.org/publ/othp69.htm).
10. Bank of England and HM Treasury. 2024. Response to the Bank of England and HM Treasury consultation paper 'The digital pound: a new form of money for households and businesses?'. Consultation response, 25 January (www.bankofengland.co.uk/paper/2024/responses-to-the-digital-pound-consultation-paper).
11. Stuart Levey. 2022. Statement by Diem CEO Stuart Levey on the sale of the Diem Group's assets to Silvergate. Press release, 31 January, Diem (www.diem.com/en-us/updates/stuart-levey-statement-diem-asset-sale/).
12. McLaughlin. The regulated internet of value. (See note 10 in chapter 9.)

13 Nick Kerigan et al. 2022. The Regulated Liability Network: digital sovereign currency. White paper, 15 November (https://regulatedliabilitynetwork.org/wp-content/uploads/2022/11/The-Regulated-Liability-Network-Whitepaper.pdf).
14 UK Finance. 2023. Regulated Liability Network: UK discovery phase. Report, 12 September (www.ukfinance.org.uk/policy-and-guidance/reports-and-publications/regulated-liability-network-uk-discovery-phase).
15 Mastercard. 2023. Unlocking the potential of digital asset innovation: building a Mastercard Multi-Token Network. White paper, July (www.mastercard.com/news/media/5zmixdjy/unlocking-the-potential-of-digital-asset-innovation-building-a-mastercard-multi-token-network-1-1.pdf).
16 Joshua Farrington. 2023. Buying a home or leasing heavy machinery with … blockchain? These innovators pitched the potential of the tech. News article, 20 September, Mastercard (www.mastercard.com/news/perspectives/2023/buying-a-home-or-leasing-heavy-machinery-with-blockchain-these-innovators-pitched-the-potential-of-the-tech/).
17 Nick Kerigan et al. The Regulated Liability Network.
18 Bank for International Settlements. Blueprint for the future monetary system.
19 Bailey. New prospects for money.
20 David Birch. 2020. *The Currency Cold War: Cash and Cryptography, Hash Rates and Hegemony.* London Publishing Partnership.

Chapter 11

1 HM Land Registry. HM Land Registry Strategy 2022+. (See note 2 in preface.)
2 Bank for International Settlements, Bank of England, Coadjute and HM Land Registry. 2023. Project Meridian: innovating transactions with synchronization. Report, 18 April (www.bis.org/publ/othp63.htm).
3 HM Land Registry. 2018. HM Land Registry to explore the benefits of blockchain. Press release, 1 October (www.gov.uk/government/news/hm-land-registry-to-explore-the-benefits-of-blockchain).
4 HM Land Registry. HM Land Registry Strategy 2022+.
5 Food and Agriculture Organization of the United Nations, United Nations Economic Commission for Europe and International

Federation of Surveyors (FIG). 2022. Digital transformation and land administration: sustainable practices from the UNECE region and beyond. FIG Publication 80 (https://unece.org/sites/default/files/2023-03/FIG%20Publication%20NO%2080%20ENG.pdf).

6 Bailey. New prospects for money. (See note 2 in chapter 10.)
7 Michael Voisin, Richard Hay and Sophia Le Vesconte. 2023. Linklaters FAQs on the UKJT legal statement on digital securities. Linklaters' *FintechLinks* blog, 15 February (www.linklaters.com/en/insights/blogs/fintechlinks/2023/february/linklaters-faqs-on-the-ukjt-legal-statement).
8 UK Finance. 2023. Unlocking the power of securities tokenisation. Report, July (www.ukfinance.org.uk/policy-and-guidance/reports-and-publications/unlocking-power-securities-tokenisation).
9 Law Commission. 2023. Digital assets: final report (https://lawcom.gov.uk/document/digital-assets-final-report/).

Chapter 12

1 Cromarty. Improving the home buying and selling process in England. (See note 3 in chapter 4.)
2 See https://content.11fs.com/reports/homebuying.
3 Zillow Group. 2023. Q4 2023 financial results: shareholder letter (https://investors.zillowgroup.com/investors/overview/default.aspx).
4 Rightmove. 2023. Rightmove Investor Day presentation (https://plc.rightmove.co.uk/content/uploads/2024/01/Rightmove_Investor_Day_24_UPDATED.pdf).
5 See https://bills.parliament.uk/bills/3430/publications.
6 UK Parliament. 2022. Data Protection and Digital Information Bill: explanatory notes (https://publications.parliament.uk/pa/bills/cbill/58-03/0143/en/220143en.pdf).
7 UK Parliament. Levelling Up Committee launches inquiry on improving the home buying and selling process. (See note 9 in chapter 2.)

Chapter 13

1 UK Parliament. Levelling Up Committee launches inquiry on improving the home buying and selling process. (See note 9 in chapter 2.)

2   HM Land Registry. HM Land Registry to explore the benefits of blockchain. (See note 3 in chapter 11.)
3   O'Reilly. What is Web 2.0? (See note 4 in chapter 3.)

Chapter 14

1   Bank for International Settlements *et al*. Project Meridian: innovating transactions with synchronization. (See note 2 in chapter 11.)
2   See www.gov.uk/guidance/land-registry-portal-official-search-of-whole-with-priority.
3   Requisitions are requests to amend missing or inaccurate information. See www.gov.uk/guidance/hm-land-registry-requisitions.

Chapter 15

1   Law Commission. 2021. The law of England and Wales can accommodate smart legal contracts, concludes Law Commission. Press release, 25 November (https://lawcom.gov.uk/the-law-of-england-and-wales-can-accommodate-smart-legal-contracts-concludes-law-commission/).
2   Law Commission. 2021. Smart legal contracts: advice to government. Report, 25 November (https://lawcom.gov.uk/project/smart-contracts/). This advice outlines the potential and challenges of implementing smart legal contracts in England and Wales, emphasizing the current legal framework's capacity to support such technology.
3   LawtechUK. 2022. Smarter contracts. Report, February (https://share-eu1.hsforms.com/1fdXebndoQgW7nOg6UC6aSAg7g8s).
4   Ibid.
5   Ibid.
6   Law Commission. Smart legal contracts.

Chapter 16

1   John McCarthy. 2007. What is artificial intelligence? Stanford University, Computer Science Department, 12 November (www-formal.stanford.edu/jmc/whatisai/).
2   See www.ibm.com/history/deep-blue.

3 See https://deepmind.google/technologies/alphago/.
4 Ashish Vaswani et al. 2017. Attention is all you need. Paper presented at the Thirty-First Conference on Neural Information Processing Systems, Long Beach, CA, USA (https://arxiv.org/abs/1706.03762).

Chapter 17

1 Michael Chui et al. 2023. The economic potential of generative AI: the next productivity frontier. Report, June, McKinsey & Company (www.mckinsey.com/capabilities/mckinsey-digital/our-insights/the-economic-potential-of-generative-AI-the-next-productivity-frontier#/).
2 Edward Felten, Manav Raj and Robert Seamans. 2021. Occupational, industry, and geographic exposure to artificial intelligence: a novel dataset and its potential uses. *Strategic Management Journal* (https://doi.org/10.1002/smj.3286).

Chapter 18

1 See https://initiatives.weforum.org/defining-and-building-the-metaverse/home.
2 To explore the business benefits of 3D tours, see https://go.matterport.com/.
3 See https://mud1.org/CMS/.
4 Prashant Kumar. 2022. The history and evolution of 3D avatars. Blog post, 23 December, Flam (https://blog.flamapp.com/the-history-and-evolution-of-3d-avatars/).
5 See www.digitalhumans.com.

Chapter 19

1 Bank of England. 2021. The Bank of England's supervision of financial market infrastructures. Report, 14 December (www.bankofengland.co.uk/-/media/boe/files/annual-report/2021/supervision-of-financial-market-infrastructures-annual-report-2021.pdf).
2 Food and Agriculture Organization of the United Nations et al. Digital transformation and land administration. (See note 5 in chapter 11.)